
PHYSICS DIVISION
General Editor: L. L. GREEN

ELEMENTS AND FORMULAE OF SPECIAL RELATIVITY

ELEMENTS AND FORMULAE OF SPECIAL RELATIVITY

BY

E. A. GUGGENHEIM, M.A., SC.D., F.R.S.
Professor of Chemistry in the University of Reading 1946–1966

THE QUEEN'S AWARD TO INDUSTRY 1966

PERGAMON PRESS

OXFORD . LONDON . EDINBURGH . NEW YORK
TORONTO . SYDNEY . PARIS . BRAUNSCHWEIG

Pergamon Press Ltd., Headington Hill Hall, Oxford
4 & 5 Fitzroy Square, London W.1
Pergamon Press (Scotland) Ltd., 2 & 3 Teviot Place, Edinburgh 1
Pergamon Press Inc., 44–01 21st Street, Long Island City, New York 11101
Pergamon of Canada, Ltd., 6 Adelaide Street East, Toronto, Ontario
Pergamon Press (Aust.) Pty. Ltd., Rushcutters Bay,
Sydney, New South Wales
Pergamon Press S.A.R.L., 24 rue des Écoles, Paris 5e
Vieweg & Sohn GmbH, Burgplatz 1, Braunschweig

First edition 1967
Library of Congress Catalog Card No. 67–28656

Printed in Great Britain by W. & G. Baird, Limited, Belfast 1

08 103505 5 (Flexicover)
08 203505 9 (hard cover)

Contents

PREFACE		vii
PRINCIPAL SYMBOLS		ix
1	Kinematics	1
2	Propagation of Light	9
3	Mechanics of Single Bodies	12
4	General Mechanics	21
5	Hydrodynamics	25
6	Thermodynamics	29
7	4-Vectors	35
8	Vector Operators	40
9	Electromagnetic Field	44
10	Electrodynamics	51
11	Statistical Mechanics	57
12	Summary of Assumptions	60
13	Historical Synopsis	61

Preface

THE theory known as special relativity was formulated by Einstein in 1905. It is concerned with the invariance of physical phenomena with respect to uniform translational velocity of the frame of reference chosen to describe the phenomena. The principle has been summarized by Pauli in the words: "The postulate of relativity implies that a uniform motion of the centre of mass of the universe relative to a closed system will be without influence on the phenomena in such a system."

The theory known as general relativity was formulated by Einstein in 1916. It is concerned with the transformation of physical laws when the frame of reference is accelerated. This theory comprises a complete description of gravitation. It has important implications for cosmology. Its mathematical tool is Riemann's tensor calculus. General relativity is an important field for mathematicians and for cosmologists, but its quantitative formulation is beyond the range of chemical physicists and indeed of most physicists.

Special relativity by contrast has important implications in chemical physics and in experimental physics. Typical examples of affected phenomena are the synchrotron, the structure of atomic spectra, the mass-defect of nuclides, and electron spin. Moreover, the mathematics required in special relativity are little more than the elements of vector calculus.

This monograph is concerned only with special relativity. It is hoped that it will be useful to chemical physicists and other not-too-theoretical physicists. No attempt has been made to treat the theory historically, but a brief historical synopsis is given in the last chapter. For full details of both special relativity and general relativity, including the history of their origins, the reader is referred to *Theory of Relativity* by W. Pauli translated from the German by G. Field (Pergamon Press, 1958). There is also an outstandingly clear account by A. Sommerfeld in his *Lectures on Theoretical Physics*, Vol. III, *Electrodynamics*, translated by E. G. Ramberg (Academic Press, 1952).

Principal Symbols

x, y, z	spatial coordinates,
t	time,
K, K′	frames of reference,
$\boldsymbol{v}$	velocity of frame K′ relative to frame K,
c	speed of light in empty space,
γ_v	$(1 - v^2/c^2)^{-\frac{1}{2}}$,
α_v	artanh (v/c),
l	length,
l_0	rest-length or proper length,
$\mathcal{v}$	volume,
$\mathcal{v}_0$	rest-volume or proper volume,
$\boldsymbol{u}$	velocity of body in frame of reference,
γ_u	$(1 - u^2/c^2)^{-\frac{1}{2}}$,
n	refractive index,
E	energy,
E_0	rest-energy or proper energy,
m	rest-mass or proper mass,
$\boldsymbol{P}$	linear momentum,
$\boldsymbol{f}$	force,
ϕ	power,
$\mathcal{L}$	Lagrangian,
$\mathcal{U}$	potential energy,
$\mathcal{H}$	Hamiltonian,
q_i	generalized coordinate,
p_i	generalized momentum,
p	pressure,

H	enthalpy,
H_0	rest-enthalpy,
S	entropy,
T	thermodynamic temperature,
F	Helmholtz function,
τ	proper time,
ω_v	arctan $(ic^{-1}v)$,
$\Box$	four-dimensional analogue of ∇,
Div	four-dimensional analogue of div,
Curl	four-dimensional analogue of curl,
$\Box^2$	four-dimensional analogue of ∇^2,
T, X, Y, Z	four-dimensional coordinates,
α_{ik}	four-dimensional transformation matrix,
F_{ik}	four-dimensional antisymmetrical vector,
ρ	electric charge density,
ψ	electric scalar potential,
$\boldsymbol{A}$	magnetic vector potential,
$\boldsymbol{\Gamma}$	electric current density 4-vector,
$\boldsymbol{\Omega}$	electromagnetic potential 4-vector,
μ_0	permeability of a vacuum,
$\boldsymbol{E}$	electric field,
$\boldsymbol{B}$	magnetic induction,
e	electric charge,
$\boldsymbol{F}$	$\boldsymbol{E} + \boldsymbol{u} \times \boldsymbol{B}$

CHAPTER 1

Kinematics

KINEMATICS is concerned with length, time, and relations between them. Lengths are measured by measuring-rods. Time is measured by clocks. The essential property of an ideal measuring-rod, for which a real rod may be a more or less good substitute, is invariance with respect to time. In other words if an ideal measuring-rod is examined at various times there should be no observable difference in it. In particular the ratio of the lengths of two measuring-rods should be the same at all times. The essential property of an ideal clock, for which a real clock such as an energized tuning-fork may be a more or less good substitute, is that its state should vary in ever repeating indistinguishable cycles. The duration of a cycle defines a unit of time. When two ideal clocks, at rest with respect to each other, in the same place, are compared the ratio of the duration of their periods should be constant.

The laws of kinematics, determined by the use of ideal measuring-rods and ideal clocks, are in the absence of a gravitational field, invariant with respect to a uniform translational velocity of the frame of reference. More precisely: there exist infinitely many reference systems moving rectilinearly and uniformly relative to one another in which phenomena occur in an identical way. These systems are called "Galilean reference systems". It is unsatisfactory that we cannot, without embarking in general relativity, give a logical reason for recognizing which set is Galilean.

The above remarks are equally relevant to the kinematics of special relativity and to pre-relativistic kinematics. The former differs from the latter in two respects. Firstly, in special relativity neither lengths measured by ideal measuring-rods nor times measured by ideal clocks are independent of motion of the measuring-rods or clocks relative to the system on which the measurements are made. Secondly, in special relativity there is a universal speed denoted by c which can be approached but never reached by any material body. The basic empirical assumption of special relativity is that the value of c is the same in all frames of reference moving at uniform velocities relative to one another. This value is 2.997925×10^8 m s^{-1}.

We now consider a frame of reference K in which the cartesian coordinates of a given point are denoted by x, y, z and the time by t. We also consider a second frame of reference K′ in which the cartesian coordinates of the same point are denoted by x', y', z' and the time by t'. We assume that the two frames of reference K and K′ are moving with constant velocities relative to each other. We denote the velocity of K′ relative to K by $\boldsymbol{v}$ and the velocity of K relative to K′ by $-\boldsymbol{v}$. We may without loss of generality choose as the x-direction the direction of $\boldsymbol{v}$. We then have $v_x = v$, $v_y = 0$, $v_z = 0$. We may also choose the origins in K and K′ so that $x = x'$, $y = y'$, $z = z'$ when $t = t' = 0$. It is convenient to introduce an auxiliary dimensionless parameter γ_v defined by

$$\gamma_v = (1 - v^2/c^2)^{\frac{1}{2}} \tag{1.1}$$

We require formulae for transformations between the two frames of reference K and K′. These transformations have to satisfy the following conditions. In the limit $v/c \to 0$ the relativistic formulae must reduce to the pre-relativistic formulae. The formulae must ensure that $v/c < 1$ always. There must be complete symmetry between the two frames of reference

K and K′. All these conditions are satisfied by the transformations

$$x' = \gamma_v^{-1}(x - vt) \tag{1.2}$$

$$y' = y \tag{1.3}$$

$$z' = z \tag{1.4}$$

$$t' = \gamma_v^{-1}(t - vx/c^2) \tag{1.5}$$

Solving these equations for x and t, we obtain

$$x = \gamma_v^{-1}(x' + vt') \tag{1.6}$$

$$y = y' \tag{1.7}$$

$$z = z' \tag{1.8}$$

$$t = \gamma_v^{-1}(t' + vx'/c^2) \tag{1.9}$$

We thus see that there is complete symmetry between the frames K and K′. Thc pre-relativistic approximation is obtained by making $c \to \infty$ so that γ_v becomes unity. We also note that when $v > c$ the value of γ_v becomes imaginary; consequently all such values of v are impossible. These transformation equations were first formulated by Lorentz in 1904 and the transformation specified by these equations is, as first suggested by Poincaré in 1905, called a Lorentz transformation.

The equations of the Lorentz transformation can be written in a more elegant form by using instead of γ_v another parameter α_v defined by

$$\tanh \alpha_v = v/c \tag{1.10}$$

We have then

$$x' = x \cosh \alpha_v - ct \sinh \alpha_v \tag{1.11}$$

$$ct' = -x \sinh \alpha_v + ct \cosh \alpha_v \tag{1.12}$$

and conversely

$$x = x' \cosh \alpha_v + ct' \sinh \alpha_v \tag{1.13}$$

$$ct = x' \sinh \alpha_v + ct' \cosh \alpha_v \tag{1.14}$$

The pre-relativistic approximations valid for $v/c \ll 1$ are obtained by letting

$$\tanh \alpha_v \to \alpha_v \ , \quad \sinh \alpha_v \to \alpha_v \ , \quad \cosh \alpha_v \to 1 \tag{1.15}$$

Let us now consider a rod AB lying along the x or x' axis at rest in the frame of reference K′. The position coordinates x'_A and x'_B of its ends are independent of t' and the length l_0 of the rod at rest is given by

$$l_0 = x'_B - x'_A \tag{1.16}$$

The length l of the rod measured in the frame of reference K at time t is defined by

$$l = x_B(t) - x_A(t) \tag{1.17}$$

According to (2) we have

$$x'_A = \gamma_v^{-1}(x_A - vt) \tag{1.18}$$

$$x'_B = \gamma_v^{-1}(x_B - vt) \tag{1.19}$$

so that

$$x'_B - x'_A = \gamma_v^{-1}(x_B - x_A) \tag{1.20}$$

From (16), (17), and (20) we deduce

$$l = \gamma_v l_0 \tag{1.21}$$

The measured length of a rod is thus less in a frame of reference K with respect to which the rod is moving than in a frame of reference K′ with respect to which the rod is at rest. This effect is called the Lorentz contraction and l_0 is called the rest-length or proper length.

Since the transverse dimensions of a body are not altered by the motion it follows that its volume V obeys the relation

$$V = \gamma_v V_0 \tag{1.22}$$

and V_0 is called the rest-volume or proper volume.

Analogously the time scale is changed by the motion. We conventionally measure time intervals from $t' = t = 0$. Consider a clock at rest at the origin of the frame K′, so that

$$x' = 0, \qquad x = vt \tag{1.23}$$

Substituting (23) into (5) we obtain

$$t' = \gamma_v^{-1} t(1 - v^2/c^2) = \gamma_v t \tag{1.24}$$

or

$$t = \gamma_v^{-1} t' \tag{1.25}$$

Thus measured in the time scale of K a moving clock will lag behind a clock at rest in K.

The time dilatation gives rise to an apparent paradox which has sometimes led to controversy and confusion. The following enunciation and resolution of the paradox is almost verbatim that of Pauli and may safely be accepted as authoritative. Consider two synchronized clocks C_1 and C_2 at rest at a point P. If one of them, say C_2, is now set in motion at time $t = 0$ and made to move with constant speed u along an arbitrary curve, reaching point P′ after time t, then it will no longer be synchronous with C_1 afterwards. On arrival at P′ it will show the time $\gamma_u t$ instead of t. The same result will hold, in particular, when P and P′ coincide, i.e. when C_2 returns to its initial position. We can neglect the effect of acceleration on the clock, so long as we are dealing with a Galilean reference system. If we take the special case where C_2 is moved along the x-axis to a point Q and then back again to P, with velocity changes at P and Q, then the effect of the acceleration will certainly be independent of t and can easily be eliminated. The paradox now lies in the following statement: Let us describe the process in terms of a reference system K*, always at rest with respect to C_2. Clock C_1 will then move relative to K* in the same way as C_2 moves relative to K. Yet, at the end of the motion, clock C_2 will have lost compared with C_1, i.e. C_1 will have gained compared with C_2. The paradox is resolved by observing that the coordinate system K* in contrast to K is not a Galilean reference system and that in such a system the effect of acceleration may not be neglected.

Having disposed of transformations of length and time we can now derive transformations of velocity. From (6), (7), (8), and (9) we obtain by differentiation keeping v constant

$$\mathrm{d}x = \gamma_v^{-1}(\mathrm{d}x' + v\,\mathrm{d}t') \tag{1.26}$$

$$\mathrm{d}y = \mathrm{d}y' \tag{1.27}$$

$$\mathrm{d}z = \mathrm{d}z' \tag{1.28}$$

$$\mathrm{d}t = \gamma_v^{-1}(\mathrm{d}t' + v\,\mathrm{d}x'/c^2) \tag{1.29}$$

We denote the velocity components as follows

$$u_x = \mathrm{d}x/\mathrm{d}t, \quad u_y = \mathrm{d}y/\mathrm{d}t, \quad u_z = \mathrm{d}z/\mathrm{d}t \tag{1.30}$$

$$u'_x = \mathrm{d}x'/\mathrm{d}t', \quad u'_y = \mathrm{d}y'/\mathrm{d}t', \quad u'_z = \mathrm{d}z'/\mathrm{d}t' \tag{1.31}$$

Dividing (26), (27), and (28) in turn by (29) we obtain

$$u_x = \frac{v + u'_x}{1 + vu'_x/c^2} \tag{1.32}$$

$$u_y = \frac{\gamma_v u'_y}{1 + vu'_x/c^2} \tag{1.33}$$

$$u_z = \frac{\gamma_v u'_z}{1 + vu'_x/c^2} \tag{1.34}$$

The pre-relativistic approximations are obtained by omitting the term vu'_x/c^2 in the denominators and replacing γ_v by unity.

If we define polar coordinates by

$$u^2 = u_x^2 + u_y^2 + u_z^2, \qquad u_x = u\cos\theta \tag{1.35}$$

$$u'^2 = u_x'^2 + u_y'^2 + u_z'^2, \qquad u'_x = u'\cos\theta' \tag{1.36}$$

we deduce

$$u = \frac{\{v^2 + 2vu'\cos\theta' + u'^2 - (vu'\sin\theta'/c)^2\}^{\frac{1}{2}}}{1 + vu'\cos\theta'/c^2} \tag{1.37}$$

$$\tan\theta = \frac{(1 - v^2/c^2)^{\frac{1}{2}}u'\sin\theta'}{v + u'\cos\theta'} \tag{1.38}$$

In the special case that $\theta' = \theta = 0$ or $u_x = u$, $u_y = 0$, $u_z = 0$ equation (32) may be derived more elegantly as follows. We consider three bodies or three frames whose relative motions are uniform and in the x-direction. We denote the velocity of the second body relative to the first by u, that of the third relative to the second by v, and that of the third relative to the first by w. We extend the definition (10) to

$$\tanh\alpha_u = u/c \tag{1.39}$$

$$\tanh\alpha_v = v/c \tag{1.40}$$

$$\tanh\alpha_w = w/c \tag{1.41}$$

Then by a geometrical argument we can deduce

$$\alpha_w = \alpha_u + \alpha_v \tag{1.42}$$

Consequently

$$\tanh\alpha_w = \frac{\tanh\alpha_u + \tanh\alpha_v}{1 + \tanh\alpha_u \tanh\alpha_v} \tag{1.43}$$

Substituting (39), (40), and (41) into (43) we obtain

$$w = \frac{u + v}{1 + uv/c^2} \tag{1.44}$$

which is equivalent to (32). We observe that when $u \to c$ then $w \to c$ regardless of the value of v.

It will be useful to have the transformation formula, from the frame of reference K′ to the frame of reference K, for γ_u defined by

$$\gamma_u = (1 - u^2/c^2)^{\frac{1}{2}} \tag{1.45}$$

Using (32), (33), and (34) we have

$$\begin{aligned}
\gamma_u^2 = 1 - \frac{u^2}{c^2} &= \left\{\left(1 + \frac{vu'_x}{c^2}\right)^2 - \left(\frac{v}{c} + \frac{u'_x}{c}\right)^2 \right.\\
&\qquad \left. - \left(1 - \frac{v^2}{c^2}\right)\left(\frac{u'^2_y}{c^2} + \frac{u'^2_z}{c^2}\right)\right\} \Big/ \left(1 + \frac{vu'_x}{c^2}\right)^2 \\
&= \left(1 - \frac{v^2}{c^2} - \frac{u'^2}{c^2} + \frac{v^2 u'^2}{c^4}\right) \Big/ \left(1 + \frac{vu'_x}{c^2}\right)^2 \\
&= \left(1 - \frac{v^2}{c^2}\right)\left(1 - \frac{u'^2}{c^2}\right) \Big/ \left(1 + \frac{vu'_x}{c^2}\right)^2
\end{aligned} \tag{1.46}$$

so that

$$\gamma_u = \frac{(1 - v^2/c^2)^{\frac{1}{2}}(1 - u'^2/c^2)^{\frac{1}{2}}}{1 + vu'_x/c^2} = \frac{\gamma_v \gamma'_u}{1 + vu'_x/c^2} \tag{1.47}$$

CHAPTER 2

Propagation of Light

IN order to discuss the propagation of light we require one further hypothesis, namely that the speed of light *in vacuo* is equal to the universal limiting speed c introduced in the preceding chapter. We recall formula (1.37) for the composition of velocities u' and v to give a resultant velocity u

$$u = \frac{\{u'^2 + v^2 + 2vu'\cos\theta' - (vu'\sin\theta'/c)^2\}^{\frac{1}{2}}}{1 + vu'\cos\theta'/c^2} \tag{2.1}$$

If we put $u' = c$ we obtain

$$u = c \tag{2.2}$$

This confirms that if the speed of light is c in one frame of reference then it is also c in any other frame of reference moving with uniform velocity relative to the first frame of reference.

Consider now a frame of reference in which a star is at rest and is observed by two observers both at the origin one at rest and the other moving with a velocity of magnitude u in the x-direction. Let the star appear to the first observer to be in a direction making an angle θ' with the x-axis and to the second observer in a direction making an angle θ with the x-axis. The relation between θ and θ' is obtained from formula (1.38)

$$\tan\theta = \frac{(1 - v^2/c^2)^{\frac{1}{2}}u'\sin\theta'}{u'\cos\theta' + v} \tag{2.3}$$

by setting $u' = c$ so that

$$\tan\theta = \frac{(1 - v^2/c^2)^{\frac{1}{2}} \sin\theta'}{\cos\theta' + v/c} = \frac{\gamma_v \sin\theta'}{\cos\theta' + v/c} \tag{2.4}$$

It follows immediately that

$$\sin\theta = \frac{\gamma_v \sin\theta'}{1 + v\cos\theta'/c} \tag{2.5}$$

$$\cos\theta = \frac{\cos\theta' + v/c}{1 + v\cos\theta'/c} \tag{2.6}$$

Formulae (4), (5), and (6) describe quantitatively the phenomenon known as the aberration of light. It must be emphasized that $\boldsymbol{v}$ is the velocity of an observer relative to the star. Any reference to a star at rest or to an observer at rest would be meaningless.

We shall next discuss the phenomenon called the Doppler effect, namely the dependence of the observed frequency ν on the relative motion $\boldsymbol{u}$ of source and observer. For the sake of simplicity we shall consider only the case when the relative motion is in the direction of the line joining the observer to the source. We may without loss of generality take this as the x-direction. We use primed and unprimed symbols to denote quantities described in frames of reference stationary with respect to the source and the observer respectively. The phase of the light received may be described by $\nu(t - x/c)$ or by $\nu'(t' - x'/c)$ and these must be identical so that

$$\nu'(t' - x'/c) = \nu(t - x/c) \tag{2.7}$$

According to (1.6) and (1.9) we have

$$x = \gamma_u^{-1}(x' + ut') \tag{2.8}$$

$$ct = \gamma_u^{-1}(ct' + ux'/c) \tag{2.9}$$

Substituting (8) and (9) into (7) we obtain

$$v'(t' - x'/c) = \gamma_u^{-1} v(t' - x'/c)(1 - u/c) \tag{2.10}$$

Consequently

$$\frac{v'}{v} = \frac{1 - u/c}{(1 - u^2/c^2)^{\frac{1}{2}}} = \left(\frac{c - u}{c + u}\right)^{\frac{1}{2}} \tag{2.11}$$

We recall that u describes the velocity of the observer relative to the source of light. There is no reference to nor meaning in absolute velocity.

We shall now consider the propagation of light through a transparent medium in motion relative to the observer. For the sake of simplicity we shall restrict ourselves to the case that the direction of the light ray coincides with that of the relative motion of observer and medium. We denote the refractive index of the medium by n so that the speed of light relative to the medium is c/n. We denote the speed of the medium relative to the observer by u. We denote the speed of the light relative to the observer by V. In this notation formula (1.44) becomes

$$V = \frac{c/n + u}{1 + u/cn} \tag{2.12}$$

Formula (12) is strictly valid only for a non-dispersive medium having n independent of the frequency.

CHAPTER 3

Mechanics of Single Bodies

A convenient basis for an elementary introduction to relativistic mechanics is provided by the laws of conservation of momentum and of energy. We retain the pre-relativistic statement of these laws: when several parts of an isolated system interact the component in each direction of the momentum of the whole system remains constant and the energy of the whole system remains constant. These laws are retained in relativistic mechanics, but momentum and energy have to be redefined so as to conform with the Lorentz transformations for kinematic quantities. At the same time the new definitions must in the limit of low velocities reduce to the pre-relativistic definitions.

In relativistic mechanics, as in pre-relativistic mechanics, every body is characterized by a quantity m called its mass. This quantity m is also called rest-mass or proper mass to emphasize that it is independent of the body's velocity. In relativistic, as in pre-relativistic, mechanics, the momentum $\boldsymbol{P}$ is directly proportional to the mass m. The requirements that the definition of $\boldsymbol{P}$ should conform with the Lorentz transformation for kinetic quantities and the requirement that in the limit of small velocities the form of $\boldsymbol{P}$ should be the same as in pre-relativistic mechanics lead to the formula

$$\boldsymbol{P} = \gamma_u^{-1} m \boldsymbol{u} \tag{3.1}$$

where γ_u is defined as previously by

$$\gamma_u = (1 - u^2/c^2)^{\frac{1}{2}} \tag{3.2}$$

We observe that the pre-relativistic formula $\boldsymbol{P} = m\boldsymbol{u}$ holds in the limit $u/c \to 0$ and $\gamma_u \to 1$. Instead of giving the rather tedious derivation of (1) we shall verify in a simple example the conservation of momentum $\boldsymbol{P}$ defined by (1).

Consider two observers A and B, the latter having a velocity $\boldsymbol{v}$ in the x-direction relative to the former. They throw towards each others spheres of equal rest-mass m and with velocities $\boldsymbol{w}$ and $-\boldsymbol{w}$ in the positive and negative y-directions such that the line of centres is in the y-direction. The x-components of the two spheres remain unchanged. From symmetry it follows that A and B will observe similar motion of their respective spheres. We shall describe the situation from the point of view of A.

The velocity $\boldsymbol{u}_{\mathrm{A}}$ of A's sphere has its x-component $u_{\mathrm{A}y}$ zero both before and after collision. Its y-component $u_{\mathrm{A}x}$ changes from w to $-w$ on collision. The magnitude u_{A} of $\boldsymbol{u}_{\mathrm{A}}$ is given by $u_{\mathrm{A}}^2 = w^2$ both before and-after collision.

The velocity $\boldsymbol{u}_{\mathrm{B}}$ of B's sphere has its x-component $u_{\mathrm{B}x}$ equal to v both before and after collision. Its y-component $u_{\mathrm{B}y}$, as observed by A, changes from $-w(1 - v^2/c^2)^{\frac{1}{2}}$ to $w(1 - v^2/c^2)^{\frac{1}{2}}$. The magnitude u_{B} of $\boldsymbol{u}_{\mathrm{B}}$, as observed by A, is given by $u_{\mathrm{B}}^2 = v^2 + w^2(1 - v^2/c^2)$ both before and after collision.

Momentum in the x-direction is obviously conserved. The changes of momentum in the y-direction are more interesting. According to the definition (1) of momentum $\boldsymbol{P}$, the y-component $P_{\mathrm{A}y}$ of the momentum of A's sphere increases by

$$-2mw\left(1 - \frac{w^2}{c^2}\right)^{-\frac{1}{2}} \tag{3.3}$$

while the y-component $P_{\mathrm{B}y}$ of the momentum of B's sphere increases by

$$2mw\left(1-\frac{v^2}{c^2}\right)^{\frac{1}{2}}\left(1-\frac{v^2}{c^2}-\frac{w^2}{c^2}+\frac{v^2w^2}{c^2}\right)^{-\frac{1}{2}}$$

$$= 2mw\left(1-\frac{w^2}{c^2}\right)^{-\frac{1}{2}} \tag{3.4}$$

Thus the change in P_{By} is equal and opposite to the change in P_{Ay} as required.

We now turn to the energy E. The requirements that the definition of E should conform with the Lorentz transformation for kinetic quantities and the requirement that in the limit of small velocities the form of E should be the same as in pre-relativistic mechanics lead to the formula

$$E = \gamma_u^{-1} mc^2 \tag{3.5}$$

apart from an arbitrary additive constant having in the present context no physical significance. In the limit $u/c \to 0$ formula (5) reduces to $E = mc^2 + \frac{1}{2}mu^2$ in agreement with the pre-relativistic formula $E = \text{const.} + \frac{1}{2}mu^2$. Instead of giving the rather tedious derivation of (5) we shall verify in a simple example the conservation of energy E defined by (5).

We do not use the previous example because in it conservation of energy is assured by symmetry alone. We now choose the example of a head-on collision between a sphere A of rest-mass m_A and a sphere B of rest-mass m_B both moving in the x-direction. We begin by considering a frame of reference in which the total momentum is zero. If we denote the velocities before collision by $\boldsymbol{u}_A$ and $\boldsymbol{u}_B$ then the velocities after collision are $-\boldsymbol{u}_A$ and $-\boldsymbol{u}_B$. The condition for zero total momentum in this frame of reference is

$$m_A u_A(1 - u_A^2/c^2)^{-\frac{1}{2}} + m_B u_B(1 - u_B^2/c^2)^{-\frac{1}{2}} = 0 \tag{3.6}$$

We now transform to a frame of reference having a velocity $-\boldsymbol{v}$ in the x-direction relative to the former frame of reference.

The velocity of A before collision is

$$(v + u_{\mathrm{A}})(1 + vu_{\mathrm{A}}/c^2)^{-1} \tag{3.7}$$

Its energy E_{A} before collision is

$$E_{\mathrm{A}} = m_{\mathrm{A}}c^2\left\{1 - \left(\frac{v/c + u_{\mathrm{A}}/c}{1 + vu_{\mathrm{A}}/c^2}\right)^2\right\}^{-1} \tag{3.8}$$

Using the identity

$$1 - \left(\frac{v/c + u_{\mathrm{A}}/c}{1 + vu_{\mathrm{A}}/c^2}\right)^2 = \left(1 - \frac{u_{\mathrm{A}}^2}{c^2}\right)\left(1 - \frac{v^2}{c^2}\right)\left(1 + \frac{vu_{\mathrm{A}}}{c^2}\right)^{-2} \tag{3.9}$$

we deduce

$$E_{\mathrm{A}} = m_{\mathrm{A}}c^2(1 - u_{\mathrm{A}}^2/c^2)^{-\frac{1}{2}}(1 - v^2/c^2)^{-\frac{1}{2}}(1 + vu_{\mathrm{A}}/c^2) \tag{3.10}$$

Similarly the energy E'_{A} of A after collision is

$$E'_{\mathrm{A}} = m_{\mathrm{A}}c^2(1 - u_{\mathrm{A}}^2/c^2)^{-\frac{1}{2}}(1 - v^2/c^2)^{-\frac{1}{2}}(1 - vu_{\mathrm{A}}/c^2) \tag{3.11}$$

Hence the increase in the energy of A is

$$E'_{\mathrm{A}} - E_{\mathrm{A}} = -2m_{\mathrm{A}}u_{\mathrm{A}}(1 - u_{\mathrm{A}}^2 c^2)^{-\frac{1}{2}}v(1 - v^2/c^2)^{-\frac{1}{2}} \tag{3.12}$$

Similarly the increase in the energy of B is

$$E'_{\mathrm{B}} - E_{\mathrm{B}} = -2m_{\mathrm{B}}u_{\mathrm{B}}(1 - u_{\mathrm{B}}^2/c^2)^{-\frac{1}{2}}v(1 - v^2/c^2)^{-\frac{1}{2}} \tag{3.13}$$

The resultant increase in the total energy is

$$E'_{\mathrm{A}} - E_{\mathrm{A}} + E'_{\mathrm{B}} - E_{\mathrm{B}} = -2\{m_{\mathrm{A}}u_{\mathrm{A}}(1 - u_{\mathrm{A}}^2/c^2)^{-\frac{1}{2}} + m_{\mathrm{B}}u_{\mathrm{B}}(1 - u_{\mathrm{B}}^2/c^2)^{-\frac{1}{2}}\}v(1 - v^2/c^2)^{-\frac{1}{2}} \tag{3.14}$$

which by virtue of (6) vanishes as required.

We can also verify conservation of momentum in this same example. The momentum P_A of A before collision is

$$P_A = m_A(v + u_A)(1 - u_A^2/c^2)^{-\frac{1}{2}}(1 - v^2/c^2)^{-\frac{1}{2}} \tag{3.15}$$

and its momentum P'_A after collision is

$$P'_A = m_A(v - u_A)(1 - u_A^2/c^2)^{-\frac{1}{2}}(1 - v^2/c^2)^{-\frac{1}{2}} \tag{3.16}$$

so that the increase in the momentum of A is

$$\begin{aligned} P'_A - P_A &= -2m_A u_A(1 - u_A^2/c^2)^{-\frac{1}{2}}(1 - v^2/c^2)^{-\frac{1}{2}} \\ &= v^{-1}(E'_A - E_A) \end{aligned} \tag{3.17}$$

Similarly the increase in the momentum of B is

$$\begin{aligned} P'_B - P_B &= -2m_B u_B(1 - u_B^2/c^2)^{-\frac{1}{2}}(1 - v^2/c^2)^{-\frac{1}{2}} \\ &= v^{-1}(E'_B - E_B) \end{aligned} \tag{3.18}$$

The resultant increase in the total momentum is

$$P'_A - P_A + P'_B - P_B = v^{-1}(E'_A - E_A + E'_B - E_B) \tag{3.19}$$

which like (14) vanishes as required.

From (1), (2), and (5) it follows that

$$E^2 - E_0^2 = c^2P^2 \tag{3.20}$$

where E_0 denotes the value of mc^2 of E when the system is at rest and may be called the rest-energy. It follows that for small values of $\boldsymbol{u}$, or of $\boldsymbol{P}$, formula (20) becomes approximately

$$E - E_0 = c^2P^2/2E_0 = P^2/2m \qquad (u \ll c) \tag{3.21}$$

which is the pre-relativistic approximation for the kinetic energy $E - E_0$. When by contrast u approaches c we note that E tends to infinity. This confirms that no material system can attain the speed c.

By use of (1) and (5) we may rewrite (2) as

$$\gamma_u^2 = 1 - u^2/c^2 = 1 - c^2P^2/E^2 = 1 - uP/E \tag{3.22}$$

We shall now derive formulae for the transformation of $\boldsymbol{P}$ and of E from a frame of reference K to a frame of reference K′ having a velocity $\boldsymbol{v}$ relative to K. We may again without loss of generality choose as the x-direction the direction of $\boldsymbol{v}$ so that $v_x = v$, $v_y = 0$, $v_z = 0$. We recall formulae (1.32), (1.33), and (1.34)

$$u_x = \frac{v + u'_x}{1 + vu'_x/c^2} \tag{3.23}$$

$$u_y = \frac{\gamma_v u'_y}{1 + vu'_x/c^2} \tag{3.24}$$

$$u_z = \frac{\gamma_v u'_z}{1 + vu'_x/c^2} \tag{3.25}$$

and formula (1.47)

$$\gamma_u = \frac{\gamma_v \gamma'_u}{1 + vu'_x/c^2} \tag{3.26}$$

We have

$$P_x = \frac{mu_x}{\gamma_u} = m\frac{v + u'_x}{1 + vu'_x/c^2}\,\frac{1 + vu'_x/c^2}{\gamma_v\gamma'_u}$$

$$= m\frac{v + u'_x}{\gamma_v\gamma'_u} = \frac{P'_x}{\gamma_v}\left(1 + \frac{v}{u'_x}\right) \tag{3.27}$$

By virtue of (1) and (5) we may rewrite (27) as

$$P_x = \gamma_v^{-1}(P'_x + vE'/c^2) \tag{3.28}$$

We also have

$$P_y = \frac{mu_y}{\gamma_u} = m\frac{\gamma_v u'_y}{1 + vu'_x/c^2}\,\frac{1 + vu'_x/c^2}{\gamma_v\gamma'_u} = \frac{mu'_y}{\gamma'_u} = P'_y \tag{3.29}$$

and similarly

$$P_z = P_z' \tag{3.30}$$

For the transformation of the energy E we have

$$E = \frac{mc^2}{\gamma_u} = \frac{mc^2(1 + vu'_x/c^2)}{\gamma_v \gamma'_u}$$

$$= \gamma_v^{-1} E' \left(1 + \frac{vu'_x}{c^2}\right) = \gamma_v^{-1}(E' + vP'_x) \qquad (3.31)$$

From (28) and (31) we deduce

$$P'_x = \gamma_v^{-1}(P_x - vE/c^2) \qquad (3.32)$$

$$E' = \gamma_v^{-1}(E - vP_x) \qquad (3.33)$$

so that there is complete symmetry between K and K′. Moreover these transformations reduce to the pre-relativistic approximations when $v/c \ll 1$. Finally we observe that the transformation formulae satisfy the required relation (20)

$$(E/c)^2 - P^2 = (E'/c)^2 - P'^2 = (E_0/c)^2 \qquad (3.34)$$

The transformations for P_x, P_y, P_z, E formally resemble the transformations given in Chapter 1 for x, y, z, t. This similarity becomes especially apparent and elegant if we use the parameter α_v defined by (1.10).

$$\tanh \alpha_v = v/c \qquad (3.35)$$

We then have

$$P'_x = P_x \cosh \alpha_v - c^{-1} E \sinh \alpha_v \qquad (3.36)$$

$$c^{-1}E' = -P_x \sinh \alpha_v + c^{-1} E \cosh \alpha_v \qquad (3.37)$$

and conversely

$$P_x = P'_x \cosh \alpha_v + c^{-1} E' \sinh \alpha_v \qquad (3.38)$$

$$c^{-1}E = P'_x \sinh \alpha_v + c^{-1} E' \cosh \alpha_v \qquad (3.39)$$

As already mentioned in an isolated system there is conservation of energy E and of momentum $\boldsymbol{P}$. If one part α of an isolated system reacts mechanically with another part β of the system then the increase in E^{α} is equal to the decrease in E^{β}. Likewise the increase in $\boldsymbol{P}^{\alpha}$ is equal to the decrease in $\boldsymbol{P}^{\beta}$. The change in $\boldsymbol{P}^{\alpha}$ or $\boldsymbol{P}^{\beta}$ per unit time is called force and is denoted by $\boldsymbol{f}$. The change in E^{α} or E^{β} per unit time is called power and is denoted by ϕ. More precisely $\boldsymbol{f}$ and ϕ are defined by

$$\boldsymbol{f} = \mathrm{d}\boldsymbol{P}/\mathrm{d}t \tag{3.40}$$

$$\phi = \mathrm{d}E/\mathrm{d}t \tag{3.41}$$

A simple relation between the force $\boldsymbol{f}$ and the power ϕ is obtained as follows. We differentiate (20) with respect to t and obtain

$$E\phi = c^2\boldsymbol{P}\boldsymbol{f} \tag{3.42}$$

or by use of (1) and (5)

$$\phi = \boldsymbol{f}\boldsymbol{u} \tag{3.43}$$

the same as in pre-relativistic theory.

We now consider the transformation of $\boldsymbol{f}$ from a frame of reference K′ to a frame of reference K relative to which K has a constant velocity with components $v_x = v$, $v_y = 0$, $v_z = 0$. We have according to (1.25)

$$\mathrm{d}t = \gamma_v^{-1}\,\mathrm{d}t' \tag{3.44}$$

Using (28) we obtain

$$f_x = \frac{\mathrm{d}P_x}{\mathrm{d}t} = \frac{\mathrm{d}}{\mathrm{d}t'}\left(P'_x + \frac{vE'}{c}\right) = f'_x + \frac{v}{c}\phi' \tag{3.45}$$

Using (29) and (30) we obtain

$$f_y = \frac{dP_y}{dt} = \gamma_v \frac{dP'_y}{dt'} = \gamma_v f'_y \tag{3.46}$$

$$f_z = \frac{dP_z}{dt} = \gamma_v \frac{dP'_z}{dt'} = \gamma_v f'_z \tag{3.47}$$

In particular if the frame of reference K′ is chosen so that in it the velocity $\boldsymbol{u}'$ is zero, then using the subscript $_0$ to denote a state of instantaneous rest, we have $\boldsymbol{u} = \boldsymbol{v}$ and $\phi'_0 = 0$ and so (45) reduces to

$$f_x = f_{0x} \tag{3.48}$$

while (46) and (47) become

$$f_y = \gamma_u f_{0y} \tag{3.49}$$

$$f_z = \gamma_u f_{0z} \tag{3.50}$$

For some purposes it is convenient to consider the force per unit volume V or force density $\boldsymbol{g}$ defined by

$$\boldsymbol{g} = \boldsymbol{f}/V \tag{3.51}$$

From the transformations of $\boldsymbol{f}$ and of V we deduce

$$g_x = \gamma_u^{-1} g_{0x} \tag{3.52}$$

$$g_y = g_{0y} \tag{3.53}$$

$$g_z = g_{0z} \tag{3.54}$$

We observe that the relations between g_x, g_y, g_z and g_{0x}, g_{0y}, g_{0z} are of the same form as

$$x = \gamma_u^{-1} x_0 \tag{3.55}$$

$$y = y_0 \tag{3.56}$$

$$z = z_0 \tag{3.57}$$

The significance of this analogy will become clear in Chapter 7.

CHAPTER 4

General Mechanics

THE preceding chapter was concerned with the mechanics of single bodies. It may be regarded as a modern version of Newton's approach. In this chapter we shall outline the more general and more powerful approach due to Lagrange and to Hamilton.

We begin by summarizing the pre-relativistic treatment. The behaviour of any mechanical system is determined by its Lagrangian $\mathscr{L}$, which is a function of the positional coordinates $x, \ldots$ and the velocity components $u_x, \ldots,$ and the external forces $f_x, \ldots$ acting on the system. For simplicity we shall consider only the case that these forces are functions of the position coordinates but are independent of the time. They may then be derived from a potential $\mathscr{U}$ by equations of the form

$$f_x = -\frac{\partial \mathscr{U}}{\partial x} \tag{4.1}$$

The principle of least action then states that

$$\delta \int_{t_1}^{t_2} \mathscr{L}\, \mathrm{d}t = 0 \tag{4.2}$$

where δ denotes variation of the coordinates $x, \ldots$ and the velocity components $u_x, \ldots$ while t_1, t_2, and the end-points of the path of integration are prescribed. By the standard method of variational calculus formula (2) leads to Lagrange's equations of motion such as

$$\frac{\mathrm{d}}{\mathrm{d}t}\left(\frac{\partial \mathscr{L}}{\partial u_x}\right) = \frac{\partial \mathscr{L}}{\partial x} \tag{4.3}$$

If we define momentum $\boldsymbol{P}$ by relations such as

$$P_x = \frac{\partial \mathscr{L}}{\partial u_x} \tag{4.4}$$

we may rewrite (3) as

$$\frac{\mathrm{d}P_x}{\mathrm{d}t} = \frac{\partial \mathscr{L}}{\partial x} \tag{4.5}$$

We now define the Hamiltonian $\mathscr{H}$, a function of the coordinates $x, \ldots$ and momenta $P_x, \ldots$, by

$$\mathscr{H}(x, \ldots, P_x, \ldots) = \sum u_x P_x - \mathscr{L} \tag{4.6}$$

If we differentiate (6) using (4) and (5) we obtain

$$\mathrm{d}\mathscr{H} = \sum u_x \mathrm{d}P_x - \sum \frac{\mathrm{d}P_x}{\mathrm{d}t}\mathrm{d}x \tag{4.7}$$

or

$$\frac{\mathrm{d}x}{\mathrm{d}t} = u_x = \frac{\partial \mathscr{H}}{\partial P_x} \tag{4.8}$$

$$\frac{\mathrm{d}P_x}{\mathrm{d}t} = -\frac{\partial \mathscr{H}}{\partial x} \tag{4.9}$$

Formulae (8) and (9) are Hamilton's equations of motion.

Since formula (2) is independent of the choice of coordinates all the formulae derived from (2) remain valid if the cartesian coordinate x is replaced by a generalized coordinate q_i, the velocity u_x is replaced by a generalized velocity $\dot{q}_i$; and the linear momentum component P_x is replaced by the generalized momentum p_i. In particular the Hamiltonian $\mathscr{H}$ is related to the Lagrangian $\mathscr{L}$ by

$$\mathscr{H} = \sum \dot{q}_i p_i - \mathscr{L} \tag{4.10}$$

and Hamilton's equations of motion become

$$\frac{\mathrm{d}q_i}{\mathrm{d}t} = \frac{\partial \mathscr{H}}{\partial p_i} \tag{4.11}$$

$$\frac{\mathrm{d}p_i}{\mathrm{d}t} = -\frac{\partial \mathscr{H}}{\partial q_i} \tag{4.12}$$

From (11) and (12) it follows that

$$\frac{\mathrm{d}\mathscr{H}}{\mathrm{d}t} = \frac{\partial \mathscr{H}}{\partial p_i}\frac{\mathrm{d}p_i}{\mathrm{d}t} + \frac{\partial \mathscr{H}}{\partial q_i}\frac{\mathrm{d}q_i}{\mathrm{d}t} = 0 \tag{4.13}$$

If we define the energy, including the potential of the forces acting on the system, as equal to the Hamiltonian, then (13) is a statement of the conservation of energy.

All the above statements, definitions, and formulae are standard in pre-relativistic mechanics. They are taken over without any change in relativistic mechanics. Only the functional forms of the Lagrangian and of the Hamiltonian are changed.

We shall not attempt to specify how the Lagrangian of a complicated system may be determined. We shall confine ourselves to the simplest example of a single body moving with a velocity $\boldsymbol{u}$ relative to a given frame of reference. Its Lagrangian has the form

$$\mathscr{L} = -\gamma_u mc^2 - \mathscr{U} \tag{4.14}$$

from which it follows that

$$P_x = \frac{\partial \mathscr{L}}{\partial u_x} = \gamma_u^{-1} mu_x \tag{4.15}$$

$$P_y = \frac{\partial \mathscr{L}}{\partial u_y} = \gamma_u^{-1} mu_y \tag{4.16}$$

$$P_z = \frac{\partial \mathscr{L}}{\partial u_z} = \gamma_u^{-1} m u_z \tag{4.17}$$

in agreement with the formulae of the preceding chapter.

For the Hamiltonian we deduce

$$\begin{aligned} \mathscr{H} &= \mathbf{uP} - \mathscr{L} \\ &= \gamma_u^{-1} m u^2 + \gamma_u m c^2 + \mathscr{U} \\ &= \gamma_u^{-1} m c^2 + \mathscr{U} \\ &= E + \mathscr{U} \end{aligned} \tag{4.18}$$

thus verifying that the Hamiltonian is equal to the energy including the potential $\mathscr{U}$ of the forces acting on the system.

CHAPTER 5

Hydrodynamics

THE formulae of Chapter 4 are strictly applicable to rigid bodies. They would not apply to a compressible fluid unless the container were included in the system. This would be inconvenient. We therefore require alternative formulae for a fluid maintained at a constant uniform pressure p.

We begin by studying the transformations of pressure. We again consider a frame of reference K′ in which the system is instantaneously at rest and which has a velocity $\boldsymbol{u}$ in the x-direction relative to another frame of reference K. As usual we use primed symbols for quantities relating to the frame K′ and unprimed symbols for quantities relating to the frame K. We recall the transformations for the co-ordinates x, y, z

$$\mathrm{d}x = \gamma_u \mathrm{d}x', \quad \mathrm{d}y = \mathrm{d}y', \quad \mathrm{d}z = \mathrm{d}z' \tag{5.1}$$

and for force components f_x, f_y, f_z

$$f_x = f'_x, \quad f_y = \gamma_u f'_y, \quad f_z = \gamma_u f'_z \tag{5.2}$$

It follows that

$$f_x/\mathrm{d}y\mathrm{d}z = f'_x/\mathrm{d}y'\mathrm{d}z' \tag{5.3}$$

$$f_y/\mathrm{d}z\mathrm{d}x = f'_y/\mathrm{d}z'\mathrm{d}x' \tag{5.4}$$

$$f_z/\mathrm{d}x\mathrm{d}y = f'_z/\mathrm{d}x'\mathrm{d}y' \tag{5.5}$$

From (3), (4), and (5) we see that in each direction the pressure p, defined as the normal force per unit area, has the same value in both reference frames. Using the subscript $_0$ to denote instantaneous rest we have then

$$p = p_0 \tag{5.6}$$

Since for a system in equilibrium p_0 is isotropic it follows that p is also isotropic.

We shall now obtain formulae for the energy E and the momentum $\boldsymbol{P}$ of a fluid at a pressure p moving with a velocity $\boldsymbol{u}$. We shall find that the final formulae are strikingly simple although their derivation is rather subtle. We recall that a body moving with velocity $\boldsymbol{u}$ has an energy $\gamma_u^{-1}E_0$ and a momentum $\gamma_u^{-1}\boldsymbol{u}E_0/c^2$ so that

$$\boldsymbol{P} = \boldsymbol{u}E/c^2 \tag{5.7}$$

It is reasonable to regard the energy as located in the moving body. We may then express (7) in the words: a transfer of energy E with a velocity $\boldsymbol{u}$ implies a momentum $\boldsymbol{u}E/c^2$. We shall assume that this statement is true for any transport of energy.

Consider a fluid of rest-mass m contained in a cylinder and by a piston at each end. We assume that the fluid is initially at rest and by suitable motion of the two pistons is accelerated to a velocity $\boldsymbol{u}$ in the direction of the axis. Then its momentum $\boldsymbol{P}$ is given by

$$\boldsymbol{P} = \gamma_u^{-1}m\boldsymbol{u} + \boldsymbol{u}pV/c^2 \tag{5.8}$$

The first term is due to the bodily flow of fluid and the second term to the flow of energy. We may rewrite (8) in the form

$$\boldsymbol{P} = \boldsymbol{u}(E + pV)/c^2 \tag{5.9}$$

If we denote the force acting on the fluid by $\boldsymbol{f}$ we have

$$\boldsymbol{f} = \frac{\mathrm{d}\boldsymbol{P}}{\mathrm{d}t} = \frac{\mathrm{d}}{\mathrm{d}t}\left(\frac{E + pV}{c^2}\boldsymbol{u}\right) \tag{5.10}$$

The rate of increase of energy of the fluid is

$$\frac{\mathrm{d}E}{\mathrm{d}t} = \boldsymbol{fu} - p\frac{\mathrm{d}V}{\mathrm{d}t} \tag{5.11}$$

where the first term is the rate at which work is done by the force which produces the acceleration and the second term is the rate at which work is done by the pressure acting on a volume decreasing according to the Lorentz contraction. Substituting (10) into (11) we obtain

$$\begin{aligned}\frac{\mathrm{d}E}{\mathrm{d}t} &= \frac{\mathrm{d}}{\mathrm{d}t}\left(\frac{E+pV}{c^2}\boldsymbol{u}\right)\boldsymbol{u} - p\frac{\mathrm{d}V}{\mathrm{d}t} \\ &= \frac{u^2}{c^2}\frac{\mathrm{d}E}{\mathrm{d}t} + \left(\frac{E+pV}{c^2}\right)\boldsymbol{u}\frac{\mathrm{d}\boldsymbol{u}}{\mathrm{d}t} + \frac{u^2}{c^2}p\frac{\mathrm{d}V}{\mathrm{d}t} - p\frac{\mathrm{d}V}{\mathrm{d}t}\end{aligned} \tag{5.12}$$

so that

$$\left(1 - \frac{u^2}{c^2}\right)\frac{\mathrm{d}(E+pV)}{\mathrm{d}t} = \frac{E+pV}{c^2}\boldsymbol{u}\frac{\mathrm{d}\boldsymbol{u}}{\mathrm{d}t} \tag{5.13}$$

or

$$\frac{1}{E+pV}\frac{\mathrm{d}(E+pV)}{\mathrm{d}t} = -\frac{\frac{1}{2}}{1-u^2/c^2}\frac{\mathrm{d}(1-u^2/c^2)}{\mathrm{d}t} \tag{5.14}$$

which on integration becomes

$$(1-u^2/c^2)^{\frac{1}{2}}(E+pV) = E_0 + pV_0 \tag{5.15}$$

or

$$E + pV = \gamma_u^{-1}(E_0 + pV_0) \tag{5.16}$$

From (9) and (16) we deduce

$$\boldsymbol{P} = \gamma_u^{-1}\boldsymbol{u}(E_0 + pV_0)/c^2 \tag{5.17}$$

Introducing the enthalpy H defined by

$$H = E + pV \tag{5.18}$$

we may rewrite (16) and (17) as

$$H = \gamma_u^{-1} H_0 \tag{5.19}$$

$$\boldsymbol{P} = \gamma_u^{-1} \boldsymbol{u} H_0 / c^2 \tag{5.20}$$

From (19) and (20) it follows that

$$H^2 - c^2 P^2 = H_0{}^2 \tag{5.21}$$

CHAPTER 6

Thermodynamics

IN previous chapters it has been tacitly assumed that every system has a constant rest-mass m and a constant rest-energy E_0, the two being related by

$$E_0 = mc^2 \tag{6.1}$$

If, however, a system undergoes a temperature change or a phase change or a chemical process or a nuclear process, there will be a change in the internal energy with a corresponding change in the rest-energy E_0. According to (1) there must also be a change in the rest-mass m. The increase ΔE_0 in the rest energy is then related to the increase Δm in the rest-mass by

$$\Delta E_0 = c^2 \Delta m \tag{6.2}$$

In a chemical process the loss of mass is undetectably small. For example, when 1 kg of H atoms combine to form H_2 molecules the energy liberated is 2.2×10^8 J. Consequently the loss of mass is $2.2 \times 10^8 \text{J}/(3.0 \times 10^8 \text{m s}^{-1})^2 = 2.4 \times 10^{-9}$ kg. In nuclear processes, by contrast the change in mass is important and accounts for nuclidic masses not having integral values. For example, if a pair of protons and a pair of neutrons are converted into an α-particle the loss of mass per kilogram is 0.0076 kg. The energy liberated is $0.0076 \text{ kg} \times (3.0 \times 10^8 \text{ m s}^{-1})^2 = 6.9 \times 10^{14}$ J. Formula (2) has been verified experimentally for nuclear processes.

The relation (2) was formulated by Einstein in a paper immediately following his paper on special relativity and this relation is usually regarded as part of the theory of special relativity. In fact the relation (2) is independent of special relativity. We have seen in previous chapters that pre-relativistic theory is an approximation to special relativity valid when $u/c \ll 1$ and that when $u = 0$ there is no difference between special relativity and pre-relativistic theory. But formula (2) relates rest-energy to rest-mass. It is not concerned with high values of u/c nor indeed with u/c at all. Formula (2) therefore belongs as much to pre-relativistic theory as to special relativity. It is a historical accident that the relation (2) was not discovered before special relativity.

To sum up: whether the condition $u/c \ll 1$ holds, so that pre-relativistic formulae are adequate, or does not hold, so that the accurate formulae of special relativity are required, any change ΔE_0 in the internal energy or rest-energy implies a change Δm in the rest-mass. The two changes are interrelated by (2).

Having clarified the physical meaning of changes in E_0 we shall now derive relativistic formulae for thermodynamics. For this purpose we require one new hypothesis, namely that the entropy S is invariant. This hypothesis may be expressed in our usual notation by

$$S = S' \qquad (6.3)$$

or by

$$S = S_0 \qquad (6.4)$$

This hypothesis is in conformity with the statistical interpretation of entropy, which depends on counting a number of states.

We begin by considering a uniform non-volatile solid at negligible pressure. When the body is at rest it has only one degree of freedom. Its rest energy E_0 and its entropy S_0 are related by

$$dE_0 = \frac{dE_0}{dS_0} dS_0 \tag{6.5}$$

When the body has a velocity $\boldsymbol{u}$ we have

$$E = \gamma_u^{-1} E_0 = \gamma_u^{-1} mc^2 \tag{6.6}$$

$$\boldsymbol{P} = \gamma_u^{-1} m\boldsymbol{u} \tag{6.7}$$

so that

$$E^2 - c^2 P^2 = E_0^2 \tag{6.8}$$

Differentiating (8) we have

$$E\,dE - c^2 \boldsymbol{P}\,d\boldsymbol{P} = E_0\,dE_0 \tag{6.9}$$

and so using (6) and (7)

$$dE - \boldsymbol{u}\,d\boldsymbol{P} = \gamma_u\,dE_0 \tag{6.10}$$

Substituting (4) and (5) into (10) we obtain

$$dE = \gamma_u \frac{\partial E_0}{\partial S_0} dS + \boldsymbol{u}\,d\boldsymbol{P} \tag{6.11}$$

We now consider two identical solid bodies α and β moving relative to each other with different but constant values of $\boldsymbol{P}$. Then the classical condition for thermal equilibrium between α and β is

$$dS^\alpha + dS^\beta = 0, \quad (dE^\alpha + dE^\beta = 0, \quad d\boldsymbol{P}^\alpha = 0, \quad d\boldsymbol{P}^\beta = 0) \tag{6.12}$$

From (11) and (12) we deduce

$$\left[\gamma_u \frac{\partial E_0}{\partial S_0}\right]^\alpha = \left[\gamma_u \frac{\partial E_0}{\partial S_0}\right]^\beta \tag{6.13}$$

Formula (13) is a complete and unambiguous statement of the condition for thermal equilibrium between two identical solid systems in relative motion. There is no need to mention temperature and indeed the property of temperature will depend on its precise definition. We may choose to define temperature T by

$$T = \gamma_u \frac{dE_0}{dS_0} \tag{6.14}$$

and the condition for thermal equilibrium then becomes

$$T^\alpha = T^\beta \tag{6.15}$$

Using the definition (14) in (11) we obtain the fundamental equation

$$dE = T dS \, \boldsymbol{u} + d\boldsymbol{P} \tag{6.16}$$

If further we define the Helmholtz function (free energy) F by

$$F = E - TS \tag{6.17}$$

and use this in (16) we obtain the second fundamental equation

$$dF = -S dT + \boldsymbol{u} \, d\boldsymbol{P} \tag{6.18}$$

We now turn to the more interesting system consisting of a fluid at a uniform pressure p which we recall is invariant with respect to uniform translational velocity. When the fluid is at rest it has two degrees of freedom. Its rest-energy E_0 and its entropy S are related by

$$dE_0 = \left(\frac{\partial E_0}{\partial S_0}\right)_V dS - p \, dV \tag{6.19}$$

where V denotes volume. Introducing the enthalpy H defined by (5.18)

$$H = E + pV \tag{6.20}$$

we may rewrite (19) as

$$dH_0 = \left(\frac{\partial H_0}{\partial S_0}\right)_p dS + V dp \tag{6.21}$$

We recall formulae (5.19), (5.20), and (5.21)

$$H = \gamma_u^{-1} H_0 \tag{6.22}$$

$$\boldsymbol{P} = \gamma^{-1} \boldsymbol{u} H_0 / c^2 \tag{6.23}$$

$$H^2 - c^2 P^2 = H_0^2 \tag{6.24}$$

Differentiating (24) we have

$$H\, dH - c^2 \boldsymbol{P}\, d\boldsymbol{P} = H_0\, dH_0 \tag{6.25}$$

and so using (22) and (23)

$$dH - \boldsymbol{u}\, d\boldsymbol{P} = \gamma_u\, dH_0 \tag{6.26}$$

Substituting (21) into (26) we obtain

$$\begin{aligned} dH &= \gamma_u \left(\frac{\partial H_0}{\partial S_0}\right)_p dS + \gamma_u V dp + \boldsymbol{u}\, d\boldsymbol{P} \\ &= \gamma_u \left(\frac{\partial H_0}{\partial S_0}\right)_p dS + V_0\, dp + \boldsymbol{u}\, d\boldsymbol{P} \end{aligned} \tag{6.27}$$

The condition for thermal equilibrium between two identical systems α and β in relative motion is

$$\begin{aligned} & dS^\alpha + dS^\beta = 0 \\ & (dH - V_0\, dp)^\alpha + (dH - V_0\, dp)^\beta = 0, \\ & d\boldsymbol{P}^\alpha = 0, \qquad d\boldsymbol{P}^\beta = 0 \end{aligned} \tag{6.28}$$

This leads to

$$\left[\gamma_u \left(\frac{\partial H_0}{\partial S_0}\right)_p\right]^\alpha = \left[\gamma_u \left(\frac{\partial H_0}{\partial S_0}\right)_p\right]^\beta \tag{6.29}$$

If we define temperature T by

$$T = \gamma_u \left(\frac{\partial H_0}{\partial S_0} \right)_p \tag{6.30}$$

equation (29) becomes

$$T^\alpha = T^\beta \tag{6.31}$$

Substituting (31) into (27) we obtain

$$\mathrm{d}H = T\,\mathrm{d}S + V_0\,\mathrm{d}p + \boldsymbol{u}\,\mathrm{d}\boldsymbol{P} \tag{6.32}$$

In all these formulae for a fluid if we put $p = 0$, so that $H = E$, we recover the formulae for a rigid body.

CHAPTER 7

4-Vectors

WHEREAS space is three-dimensional and time is one-dimensional, space and time together constitute a four-dimensional *world.* A body is at each instant described by four *world coordinates* x, y, z, t and its progress by the four differential coordinates $\mathrm{d}x$, $\mathrm{d}y$, $\mathrm{d}z$, $\mathrm{d}t$. Transformations of coordinates in pre-relativistic theory, commonly called Galilean transformations, were required to satisfy the conditions that $(\mathrm{d}x)^2 + (\mathrm{d}y)^2 + (\mathrm{d}z)^2$ should be invariant and that $\mathrm{d}t$ should also be invariant. Relativity theory imposes the less stringent condition.

$$(c\,\mathrm{d}t)^2 - (\mathrm{d}x)^2 - (\mathrm{d}y)^2 - (\mathrm{d}z)^2 = (c\,\mathrm{d}\tau)^2$$
$$(\mathrm{d}\tau \text{ invariant}) \qquad (7.1)$$

A Lorentz transformation is a transformation of coordinates satisfying condition (1).

Since two successive Lorentz transformations are equivalent to another single Lorentz transformation and since every Lorentz transformation has its inverse transformation the totality of all possible Lorentz transformations forms a group. This transformation group is defined by the condition (1). The group includes as a sub-group the pre-relativistic transformations mentioned in the preceding paragraph. Another sub-group, occurring frequently in earlier chapters, is defined by the simultaneous invariance of $(c\,\mathrm{d}t)^2 - (\mathrm{d}x)^2$, of $\mathrm{d}y$, and of $\mathrm{d}z$. The condition (1) is except for the minus signs the same as the condition for the transformation of a four-dimensional vector. This was pointed out by Minkowski. We therefore call $c\mathrm{d}t$, $\mathrm{d}x$, $\mathrm{d}y$, $\mathrm{d}z$ the components of a Minkowski-vector or M-vector. We also call $c\,\mathrm{d}\tau$ the *length* of the M-vector, observing that the square of this *length* may be negative.

The principle of special relativity requires that all the physical laws covered by this principle can be expressed in

TABLE 7.1 IMPORTANT M-VECTORS

	Components				Length
1.	$\mathrm{d}x$	$\mathrm{d}y$	$\mathrm{d}z$	$c\,\mathrm{d}t$	$c\,\mathrm{d}\tau$
2.	$\gamma_u^{-1}u_x$	$\gamma_u^{-1}u_y$	$\gamma_u^{-1}u_z$	$\gamma_u^{-1}c$	c
3.	P_x	P_y	P_z	$c^{-1}E$	$c^{-1}E_0$
4.	$\gamma_u^{-1}f_x$	$\gamma_u^{-1}f_y$	$\gamma_u^{-1}f_z$	$\gamma_u^{-1}c^{-1}\phi$	$-if_0$
5.	$g_x = V^{-1}f_x$	$g_y = V^{-1}f_y$	$g_z = V^{-1}f_z$	$c^{-1}V^{-1}\phi$	$-ig_0 = if_0V_0^{-1}$

terms of M-vectors. We shall now verify this for the several relations specified in the earlier chapters, The facts are summarized in Table 7.1. The first four columns give the four components of an M-vector and the fifth column gives its length and this length must be invariant if the quantity being considered is indeed an M-vector.

The M-vector in the first row is the one already specified by (1). We observe that for a frame of reference in which a clock is at rest $\mathrm{d}t = \mathrm{d}\tau$. We may therefore call τ the rest-time or local time or proper time. The last name is the most used.

If we divide the M-vector in the first row by the invariant quantity $\mathrm{d}\tau = \gamma_u\,\mathrm{d}t$ we obtain the M-vector of the second row. We recall the definition of velocity $u_x = \mathrm{d}x/\mathrm{d}t$, $u_y = \mathrm{d}y/\mathrm{d}t$,

$u_z = dz/dt$ and we note that the corresponding factor in the t column is c. If we multiply the M-vector in the second row by the rest mass m we obtain the momentum-energy M-vector in the third row. We observe that

$$\left(\frac{E}{c}\right)^2 - P_x^2 - P_y^2 - P_z^2 = \left(\frac{E}{c}\right)^2 - P^2 = \left(\frac{E_0}{c}\right)^2 \qquad (7.2)$$

If we differentiate the momentum-energy M-vector with respect to the invariant quantity $\tau = \gamma_u t$ we obtain the vector in the fourth row. As previously f denotes force and ϕ denotes power, so that

$$\phi = \boldsymbol{fu} = f_x u_x + f_y u_y + f_z u_z \qquad (7.3)$$

We recall the dependence of volume V on velocity $\boldsymbol{u}$

$$V = \gamma_u V_0 \qquad (7.4)$$

Using (4) we derive from the M-vector in the fourth row the force-density and power-density vectors of the fifth row. It is important to remember that force-density, but not force, forms three components of an M-vector. This fact recurs in the next chapter.

We recall that the progress of a body through the world is described by the M-vector whose components $c\mathrm{d}t$, $\mathrm{d}x$, $\mathrm{d}y$, $\mathrm{d}z$ satisfy the equation

$$(c\,\mathrm{d}t)^2 - (\mathrm{d}x)^2 - (\mathrm{d}y)^2 - (\mathrm{d}z)^2 = (c\,\mathrm{d}\tau)^2$$
$$(\mathrm{d}\tau \text{ invariant}) \qquad (7.5)$$

When $(\mathrm{d}\tau)^2 > 0$ this world vector is called time-like because it is possible to choose a frame of reference in which $\mathrm{d}x$, $\mathrm{d}y$, $\mathrm{d}z$ vanish so that $\mathrm{d}\tau$ decomes identical with $\mathrm{d}t$. When on the contrary $(\mathrm{d}\tau)^2 < 0$ we set

$$(c\,\mathrm{d}\tau)^2 = -(\mathrm{d}\lambda)^2 \qquad (7.6)$$

and we rewrite (5) as

$$(c\,\mathrm{d}t)^2-(\mathrm{d}x)^2-(\mathrm{d}y)^2-(\mathrm{d}z)^2=-(\mathrm{d}\lambda)^2$$
$$(\mathrm{d}\lambda \text{ invariant}) \tag{7.7}$$

In this case it is possible to choose a frame of reference in which $\mathrm{d}\tau$, $\mathrm{d}y$, $\mathrm{d}z$ vanish so that $\mathrm{d}\lambda$ becomes identical with $\mathrm{d}x$. Such a world vector is called space-like.

There is a striking gain in elegance and sometimes a gain in convenience if the M-vectors, with real components, are represented by cartesian 4-vectors with imaginary components. This is illustrated by three examples in Table 7.2.

TABLE 7.2 IMPORTANT 4-VECTORS

	Components				Length
1.	$c\mathrm{d}t$	$-i\mathrm{d}x$	$-i\mathrm{d}y$	$-i\mathrm{d}z$	$c\mathrm{d}\tau$
3.	$c^{-1}E$	$-iP_x$	$-iP_y$	$-iP_z$	$c^{-1}E_0$
5.	$c^{-1}V^{-1}\phi$	$-iV^{-1}f_x$	$-iV^{-1}f_y$	$-iV^{-1}f_z$	$-if_0V_0^{-1}$
		$=-ig_x$	$=-ig_y$	$=-ig_z$	$=-ig_0$

Since these are cartesian vectors the sum of the squares of the four components is equal to the square of the invariant length of the 4-vector.

At this stage it is instructive to turn back to formulae (1.10)–(1.14) which we now rewrite as follows

$$\tan\omega_v=ivc^{-1}=i\tanh\alpha_v,\qquad \omega_v=i\alpha_v \tag{7.8}$$

$$ct'=ct\cos\omega_v+ix\sin\omega_v \tag{7.9}$$

$$-ix'=ct\sin\omega_v-ix\cos\omega_v \tag{7.10}$$

and conversely

$$ct = ct' \cos \omega_v \quad - ix' \sin \omega_v \tag{7.11}$$

$$-ix = -ct' \sin \omega_v - ix' \cos \omega_v \tag{7.12}$$

From these relations we see that the transformations between the frames of reference K and K′, having a relative velocity v along the x-axis, consist in rotations in the $(ct, -ix)$ plane through an imaginary angle ω_v defined by (8).

CHAPTER 8

Vector Operators

VECTOR operators have, in common with vectors, invariant properties independent of the choice of coordinate system. In this respect 4-vectors are similar to three-dimensional vectors.

By analogy with

$$\nabla A = \operatorname{div} A = \frac{\partial A_x}{\partial x} + \frac{\partial A_y}{\partial y} + \frac{\partial A_z}{\partial z} \tag{8.1}$$

we write

$$\Box A = \operatorname{Div} A = \frac{\partial A_w}{\partial w} + \frac{\partial A_x}{\partial x} + \frac{\partial A_y}{\partial y} + \frac{\partial A_z}{\partial z} \tag{8.2}$$

Also by analogy with

$$\nabla^2 A = \left(\frac{\partial^2}{\partial x^2} + \frac{\partial^2}{\partial y^2} + \frac{\partial^2}{\partial z^2}\right) A \tag{8.3}$$

we write

$$\Box^2 A = \left(\frac{\partial^2}{\partial w^2} + \frac{\partial^2}{\partial x^2} + \frac{\partial^2}{\partial y^2} + \frac{\partial^2}{\partial w^2}\right) A \tag{8.4}$$

There also exists a close analogy between the curl operator in three dimensions and a Curl operator in four dimensions, but to appreciate this it is necessary to improve the usual notation in three dimensions. We accordingly define the components of curl in three dimensions by

$$(\text{curl}\,\boldsymbol{A})_{xy} = \frac{\partial A_y}{\partial x} - \frac{\partial A_x}{\partial y} = -(\text{curl}\,\boldsymbol{A})_{yx} \tag{8.5}$$

$$(\text{curl}\,\boldsymbol{A})_{yz} = \frac{\partial A_z}{\partial y} - \frac{\partial A_y}{\partial z} = -(\text{curl}\,\boldsymbol{A})_{zy} \tag{8.6}$$

$$(\text{curl}\,\boldsymbol{A})_{zx} = \frac{\partial A_x}{\partial z} - \frac{\partial A_z}{\partial x} = -(\text{curl}\,\boldsymbol{A})_{xz} \tag{8.7}$$

Analogously we define the components of Curl in four dimensions by

$$(\text{curl}\,\boldsymbol{A})_{wx} = \frac{\partial A_x}{\partial w} - \frac{\partial A_w}{\partial x} = -(\text{curl}\,\boldsymbol{A})_{xw} \tag{8.8}$$

$$(\text{curl}\,\boldsymbol{A})_{wy} = \frac{\partial A_y}{\partial w} - \frac{\partial A_w}{\partial y} = -(\text{curl}\,\boldsymbol{A})_{yw} \tag{8.9}$$

$$(\text{curl}\,\boldsymbol{A})_{wz} = \frac{\partial A_z}{\partial w} - \frac{\partial A_w}{\partial z} = -(\text{curl}\,\boldsymbol{A})_{zw} \tag{8.10}$$

$$(\text{curl}\,\boldsymbol{A})_{xy} = \frac{\partial A_y}{\partial x} - \frac{\partial A_x}{\partial y} = -(\text{curl}\,\boldsymbol{A})_{yx} \tag{8.11}$$

$$(\text{curl}\,\boldsymbol{A})_{yz} = \frac{\partial A_z}{\partial y} - \frac{\partial A_y}{\partial z} = -(\text{curl}\,\boldsymbol{A})_{zy} \tag{8.12}$$

$$(\text{curl}\,\boldsymbol{A})_{zx} = \frac{\partial A_x}{\partial z} - \frac{\partial A_z}{\partial x} = -(\text{curl}\,\boldsymbol{A})_{xz} \tag{8.13}$$

It is evident from these definitions that neither curl $\boldsymbol{A}$ nor Curl $\boldsymbol{A}$ is a vector. Both are in fact antisymmetrical tensors or skew-symmetrical tensors. We note that curl $\boldsymbol{A}$ has six non-zero components and Curl $\boldsymbol{A}$ has twelve non-zero components. If we disregard the positive or negative signs then curl $\boldsymbol{A}$ has three non-zero components and Curl $\boldsymbol{A}$ has six. It is easy to generalize this. In n dimensions there are $\frac{1}{2}n(n-1)$ non-zero components apart from difference of sign. It is an accident that only when $n = 3$ we have $\frac{1}{2}n(n-1) = n$. This explains why in three dimensions, and only in three dimensions, it is possible to replace the universally correct notation $(\text{curl } A)_{yz}$, $(\text{curl } A)_{zx}$, $(\text{curl } A)_{xy}$ by the misleading notation $(\text{curl } A)_x$, $(\text{curl } A)_y$, $(\text{curl } A)_z$. That is to say in three dimensions and only in three dimensions it is possible to represent an antisymmetrical tensor by a vector.

We have already mentioned that in four dimensions the Curl of a vector is not a vector but is an antisymmetric tensor. Since we shall later require to transform these antisymmetric tensors from one frame of reference to another, we now give the formulae for these transformations. For the present purpose only we denote the four cartesian coordinates by T, X, Y, Z. The transformation of a vector A is conveniently described by the matrix α displayed in Table 8.1. We have

$$A'_T = \alpha_{TT}A_T + \alpha_{TX}A_X + \alpha_{TY}A_Y + \alpha_{TZ}A_Z \tag{8.14}$$

$$A'_X = \alpha_{XT}A_T + \alpha_{XX}A_X + \alpha_{XY}A_Y + \alpha_{XZ}A_Z \tag{8.15}$$

$$A'_Y = \alpha_{YT}A_T + \alpha_{YX}A_X + \alpha_{YY}A_Y + \alpha_{YZ}A_Z \tag{8.16}$$

$$A'_Z = \alpha_{ZT}A_T + \alpha_{ZX}A_X + \alpha_{ZY}A_Y + \alpha_{ZZ}A_Z \tag{8.17}$$

TABLE 8.1

	T	X	Y	Z
T'	α_{TT}	α_{TX}	α_{TY}	α_{TZ}
X'	α_{XT}	α_{XX}	α_{XY}	α_{XZ}
Y'	α_{YT}	α_{YX}	α_{YY}	α_{YZ}
Z'	α_{ZT}	α_{ZX}	α_{ZY}	α_{ZZ}

A tensor F is a quantity whose components F_{nn} and F_{nm} behave in a transformation like the squares and products of the components of a vector. We thus have

$$F'_{nm} = \sum_i \sum_k \alpha_{ni} \alpha_{mk} F_{ik} \tag{8.18}$$

If the tensor is antisymmetrical (18) may be written as

$$F'_{nm} = \sum_{i,k} (\alpha_{ni} \alpha_{mk} - \alpha_{mi} \alpha_{nk}) F_{ik} \tag{8.19}$$

These formulae will be used in Chapters 9 and 10.

CHAPTER 9

Electromagnetic Field

In this chapter we shall obtain the equations for an electromagnetic field in a vacuum. We shall verify that the equations of Maxwell are invariant with respect to a Lorentz transformation. This was first proved by Lorentz in 1904 and independently by Einstein in 1905, but the most elegant and most powerful treatment was given by Minkowski in 1907. The treatment can be expressed either by means of M-vectors with real components or by means of cartesian 4-vectors with imaginary components. We have chosen the second alternative.

We begin by referring to the several 4-vectors displayed in Table 9.1. The chosen coordinates are specified by the first 4-vector. The second and third vectors are operators analogous to the three-dimensional div and ∇^2 respectively. The fourth vector, denoted by $\boldsymbol{\Gamma}$ has components related to the electric charge density ρ and the electric current density $\boldsymbol{J} = \rho \boldsymbol{u}$. The classical equation of continuity

$$\frac{\partial \rho}{\partial t} + \operatorname{div} \boldsymbol{J} = 0 \tag{9.1}$$

is equivalent to

$$\operatorname{Div} \boldsymbol{\Gamma} = 0 \tag{9.2}$$

We now postulate the existence of the fifth 4-vector $\boldsymbol{\Omega}$ satisfying the equation

$$\operatorname{Div} \boldsymbol{\Omega} = 0 \tag{9.3}$$

TABLE 9.1

	4-vector or operator	Components			
1.	$cd\tau$	cdt	$-idx$	$-idy$	$-idz$
2.	$\Box$ or Div	$\frac{1}{c}\frac{\partial}{\partial t}$	$i\frac{\partial}{\partial x}$	$i\frac{\partial}{\partial y}$	$i\frac{\partial}{\partial z}$
3.	$\Box^2$	$\frac{1}{c^2}\frac{\partial^2}{\partial t^2}$	$-\frac{\partial^2}{\partial x^2}$	$-\frac{\partial^2}{\partial y^2}$	$-\frac{\partial^2}{\partial z^2}$
4.	$\boldsymbol{\Gamma}$	$c\rho$	$-iJ_x$	$-iJ_y$	$-iJ_z$
5.	$\boldsymbol{\Omega}$	$c^{-1}\psi$	$-iA_x$	$-iA_y$	$-iA_z$

and related to $\boldsymbol{\Gamma}$ by the equation

$$\Box^2\boldsymbol{\Omega} = \mu_0\boldsymbol{\Gamma} \tag{9.4}$$

where μ_0 is the permeability of a vacuum. We shall show that ψ has all the properties of the scalar electric potential and A_x, A_y, A_z have all the properties of the components of the magnetic vector potential $\boldsymbol{A}$.

We rewrite (3) in the expanded form

$$\frac{1}{c^2}\frac{\partial\psi}{\partial t} + \frac{\partial A_x}{\partial x} + \frac{\partial A_y}{\partial y} + \frac{\partial A_z}{\partial z} = 0 \tag{9.5}$$

or

$$\frac{1}{c^2}\frac{\partial\psi}{\partial t} + \operatorname{div}\boldsymbol{A} = 0 \tag{9.6}$$

We also rewrite (4) in the expanded form

$$\frac{1}{c^2}\frac{\partial^2\psi}{\partial t^2} - \frac{\partial^2\psi}{\partial x^2} - \frac{\partial^2\psi}{\partial y^2} - \frac{\partial^2\psi}{\partial z^2} = \mu_0 c^2\rho \tag{9.7}$$

$$\frac{1}{c^2}\frac{\partial^2 A_x}{\partial t^2} - \frac{\partial^2 A_x}{\partial x^2} - \frac{\partial^2 A_x}{\partial y^2} - \frac{\partial^2 A_x}{\partial z^2} = \mu_0 J_x \tag{9.8}$$

$$\frac{1}{c^2}\frac{\partial^2 A_y}{\partial t^2} - \frac{\partial^2 A_y}{\partial x^2} - \frac{\partial^2 A_y}{\partial y^2} - \frac{\partial^2 A_y}{\partial z^2} = \mu_0 J_y \tag{9.9}$$

$$\frac{1}{c^2}\frac{\partial^2 A_z}{\partial t^2} - \frac{\partial^2 A_z}{\partial x^2} - \frac{\partial^2 A_z}{\partial y^2} - \frac{\partial^2 A_z}{\partial z^2} = \mu_0 J_z \tag{9.10}$$

or

$$\frac{1}{c^2}\frac{\partial^2 \psi}{\partial t^2} - \nabla^2 \psi = \mu_0 c^2 \rho \tag{9.11}$$

$$\frac{1}{c^2}\frac{\partial^2 \boldsymbol{A}}{\partial t^2} - \nabla^2 \boldsymbol{A} = \mu_0 \boldsymbol{J} \tag{9.12}$$

We now construct Curl $\boldsymbol{\Omega}$ and call three of the components $ic^{-1}E_x$, $ic^{-1}E_y$, $ic^{-1}E_z$ as follows.

$$\frac{1}{c}\frac{\partial}{\partial t}(-iA_x) - i\frac{\partial}{\partial x}\left(\frac{1}{c}\psi\right) = \frac{i}{c}E_x \tag{9.13}$$

$$\frac{1}{c}\frac{\partial}{\partial t}(-iA_y) - i\frac{\partial}{\partial y}\left(\frac{1}{c}\psi\right) = \frac{i}{c}E_y \tag{9.14}$$

$$\frac{1}{c}\frac{\partial}{\partial t}(-iA_z) - i\frac{\partial}{\partial z}\left(\frac{1}{c}\psi\right) = \frac{i}{c}E_z \tag{9.15}$$

If we regard E_x, E_y, E_z as the components of a three-dimensional vector $\boldsymbol{E}$ we may combine the last three equations into

$$\boldsymbol{E} = -\frac{\partial \boldsymbol{A}}{\partial t} - \operatorname{grad} \psi \tag{9.16}$$

Three other components of Curl $\boldsymbol{\Omega}$ are denoted by $-B_x$, $-B_y$, $-B_z$ as follows.

$$i\frac{\partial}{\partial y}(-iA_z) - i\frac{\partial}{\partial z}(-iA_y) = -B_x \tag{9.17}$$

$$i\frac{\partial}{\partial z}(-iA_x) - i\frac{\partial}{\partial x}(-iA_z) = -B_y \tag{9.18}$$

$$i\frac{\partial}{\partial x}(-iA_y) - i\frac{\partial}{\partial y}(-iA_x) = -B_z \tag{9.19}$$

If we treat B_x, B_y, B_z as the components of a three-dimensional vector $\boldsymbol{B}$ we may combine the last three equations into

$$\boldsymbol{B} = \text{curl}\, \boldsymbol{A} \tag{9.20}$$

From (16) we deduce, using (20),

$$\text{curl}\, \boldsymbol{E} = -\frac{\partial \boldsymbol{B}}{\partial t} \tag{9.21}$$

From (20) we also deduce

$$\text{div}\, \boldsymbol{B} = 0 \tag{9.22}$$

From (16) we deduce, using (6)

$$\text{div}\, \boldsymbol{E} = -\frac{\partial}{\partial t}(\text{div}\, \boldsymbol{A}) - \nabla^2 \psi = \frac{1}{c^2}\frac{\partial^2 \psi}{\partial t^2} - \nabla^2 \psi \tag{9.23}$$

and using (11)

$$\frac{1}{\mu_0 c^2}\text{div}\, \boldsymbol{E} = \rho \tag{9.24}$$

From (20) we deduce, using (6) and (12)

$$\begin{aligned}\text{curl}\, \boldsymbol{B} &= \text{curl}\,\text{curl}\, \boldsymbol{A} = \text{grad}\,\text{div}\, \boldsymbol{A} - \nabla^2 \boldsymbol{A} \\ &= -\frac{1}{c^2}\frac{\partial}{\partial t}(\text{grad}\, \psi) - \frac{1}{c^2}\frac{\partial^2 \boldsymbol{A}}{\partial t^2} + \mu_0 J \\ &= -\frac{1}{c^2}\frac{\partial}{\partial t}\left(\text{grad}\, \psi + \frac{\partial \boldsymbol{A}}{\partial t}\right) + \mu_0 J \end{aligned} \tag{9.25}$$

and then using (16)

$$\frac{1}{\mu_0}\operatorname{curl}\boldsymbol{B} = \frac{1}{\mu_0 c^2}\frac{\partial E}{\partial t} + \boldsymbol{J} \tag{9.26}$$

We recognize (21), (22), (24), (26) together with (1) as Maxwell's equations with $\boldsymbol{E}$ denoting the electric field, $\boldsymbol{B}$ the magnetic induction, μ_0 the permeability of a vacuum and $(\mu_0 c^2)^{-1}$ the permittivity of a vacuum. From (16) and (20) it follows that ψ is indeed the electric scalar potential and $\boldsymbol{A}$ is indeed the magnetic vector potential.

We have now shown that Maxwell's equations and all the subsidiary equations tied to them remain unaltered by a change in velocity of the frame of reference provided that $\boldsymbol{\Omega}$ constructed from ψ and $\boldsymbol{A}$, as shown in Table 9.1, is a cartesian 4-vector and as such obeys the transformation rules for 4-vectors.

In particular the transformation from a frame of reference K′ to a frame of reference K, when the velocity of K′ relative to K is $\boldsymbol{v}$ along the x-axis, becomes

$$c^{-1}\psi' = c^{-1}\psi\cos\omega_v + iA_x\sin\omega_v \tag{9.27}$$

$$-iA'_x = c^{-1}\psi\sin\omega_v - iA_x\cos\omega_v \tag{9.28}$$

where in accordance with (7.8)

$$\tan\omega_v = ic^{-1}v \tag{9.29}$$

Conversely

$$c^{-1}\psi = c^{-1}\psi'\cos\omega_v - iA'_x\sin\omega_v \tag{9.30}$$

$$-iA_x = c^{-1}\psi'\sin\omega_v - iA'_x\cos\omega_v \tag{9.31}$$

together with

$$A'_y = A_y \tag{9.32}$$

$$A'_z = A_z \tag{9.33}$$

We shall conclude this chapter by deriving the transformation formulae for $\boldsymbol{E}$ and $\boldsymbol{B}$. The matrix α shown in Table 8.1 now takes the form shown in Table 9.2 wherein in accordance with (29)

$$\sin \omega_v = \gamma_v^{-1} i c^{-1} v \tag{9.34}$$

$$\cos \omega_v = \gamma_v^{-1} \tag{9.35}$$

Using formula (8.19) we obtain

$$(\text{Curl}\,\boldsymbol{\Omega}')_{TX} = (\text{Curl}\,\boldsymbol{\Omega})_{TX} \tag{9.36}$$

$$(\text{Curl}\,\boldsymbol{\Omega}')_{TY} = \cos \omega_v (\text{Curl}\,\boldsymbol{\Omega})_{TY} + \sin \omega_v (\text{Curl}\,\boldsymbol{\Omega})_{XY} \tag{9.37}$$

$$(\text{Curl}\,\boldsymbol{\Omega}')_{TZ} = \cos \omega_v (\text{Curl}\,\boldsymbol{\Omega})_{TZ} - \sin \omega_v (\text{Curl}\,\boldsymbol{\Omega})_{ZX} \tag{9.38}$$

$$(\text{Curl}\,\boldsymbol{\Omega}')_{XY} = \cos \omega_v (\text{Curl}\boldsymbol{\Omega})_{XY} + \sin \omega_v (\text{Curl}\,\boldsymbol{\Omega})_{YT} \tag{9.39}$$

$$(\text{Curl}\,\boldsymbol{\Omega}')_{YZ} = (\text{Curl}\,\boldsymbol{\Omega})_{YZ} \tag{9.40}$$

$$(\text{Curl}\,\boldsymbol{\Omega}')_{ZX} = \cos \omega_v (\text{Curl}\boldsymbol{\Omega})_{ZX} - \sin \omega_v (\text{Curl}\boldsymbol{\Omega})_{ZT} \tag{9.41}$$

We recall the definitions of $\boldsymbol{E}$ and of $\boldsymbol{B}$ in (13)–(19). Using these we rewrite (36), (37), and (38) as

$$E'_x = E_x \tag{9.42}$$

$$E'_y = \gamma_v^{-1} E_y - \gamma_v^{-1} v B_z \tag{9.43}$$

$$E'_z = \gamma_v^{-1} E_z + \gamma_v^{-1} v B_y \tag{9.44}$$

Similarly we rewrite (39), (40), and (41) as

$$B'_x = B_x \tag{9.45}$$

$$B'_y = \gamma_v^{-1} B_y + \gamma_v^{-1} c^{-2} v E_z \tag{9.46}$$

$$B'_z = \gamma_v^{-1} B_z - \gamma_v^{-1} c^{-2} v E_y \tag{9.47}$$

Using the fact that $v_y = 0$, $v_z = 0$ we may rewrite (42), (43), and (44) as

$$E'_x = (\boldsymbol{E} + \boldsymbol{v} \times \boldsymbol{B})_x \tag{9.48}$$

$$E'_y = \gamma_v^{-1}(\boldsymbol{E} + \boldsymbol{v} \times \boldsymbol{B})_y \tag{9.49}$$

$$E'_z = \gamma_v^{-1}(\boldsymbol{E} + \boldsymbol{v} \times \boldsymbol{B})_z \tag{9.50}$$

TABLE 9.2

	T $= ct$	X $= -ix$	Y $= -iy$	Z $= -iz$
$T' = ct'$	$\cos \omega_v$	$-\sin \omega_v$	0	0
$X' = -ix'$	$\sin \omega_v$	$\cos \omega_v$	0	0
$Y' = -iy'$	0	0	1	0
$Z' = -iz'$	0	0	0	1

Likewise we may rewrite (45), (46), and (47) as

$$B'_x = (\boldsymbol{B} - c^{-2}\boldsymbol{v} \times \boldsymbol{E})_x \tag{9.51}$$

$$B'_y = \gamma_v^{-1}(\boldsymbol{B} - c^{-2}\boldsymbol{v} \times \boldsymbol{E})_y \tag{9.52}$$

$$B'_z = \gamma_v^{-1}(\boldsymbol{B} - c^{-2}\boldsymbol{v} \times \boldsymbol{E})_z \tag{9.53}$$

Formulae (48), (49), and (50) will play an important part in the next chapter.

CHAPTER 10

Electrodynamics

THE preceding chapter gave a comprehensive account of the several quantities ψ, $\boldsymbol{A}$, $\boldsymbol{E}$, $\boldsymbol{B}$ which describe the electromagnetic field, of the relations between them, and of their transformations from one reference frame to another. Nothing was said about the physical significance of these quantities. The present chapter is concerned with this question and particularly with the determination of the force acting on an electrically charged test-body assumed small enough for the external field to be regarded as uniform throughout this test-body.

In regarding formula (9.1) or (9.2) to be an expression of the continuity of electric charge it was tacitly assumed that the electric charge is unaltered by motion. We now make this an explicit assumption. It is a reasonable assumption since electric charge can be measured by the integral number of elementary charges. Moreover, as pointed out by Pauli, otherwise the neutral character of an atom could be upset by the mere motion of the electrons which it contains. We further assume that the force on an electrically charged test-body in a given field is directly proportional to its charge e. Finally we assume that the force on a charge e at rest is $e\boldsymbol{E}$.

We accordingly write for the force $\boldsymbol{f}$

$$\boldsymbol{f} = e\boldsymbol{F} \qquad (10.1)$$

and tentatively assume Lorentz's pre-relativistic formula

$$\boldsymbol{F} = \boldsymbol{E} + \boldsymbol{u} \times \boldsymbol{B} \tag{10.2}$$

where $\boldsymbol{u}$ is the velocity of the charged body. We see at once that for a charge at rest this leads to the required relation $\boldsymbol{f} = e\boldsymbol{E}$.

We recall formulae (9.48), (9.49), and (9.50). We now assume that the charged body is at rest in the reference frame K′ and has a velocity $\boldsymbol{u} = \boldsymbol{v}$ in the reference frame K. Using (2) we may then rewrite these formulae as

$$F_x = F_{0x} \tag{10.3}$$

$$F_y = \gamma_u F_{0y} \tag{10.4}$$

$$F_z = \gamma_u F_{0z} \tag{10.5}$$

where as usual the subscript $_0$ denotes a state of rest.

When we compare (3), (4), and (5) with (3.29), (3.30), and (3.31) respectively we see that $\boldsymbol{F}$, and consequently also $e\boldsymbol{F}$, transforms precisely as a force. Thus $\boldsymbol{F}$ defined by (2) has all the properties required by the theory.

The force per unit volume or force density $\boldsymbol{g} = \boldsymbol{f}/V$ is equal to $\rho\boldsymbol{F}$ and the power density ϕ/V is equal to $\rho\boldsymbol{F}\boldsymbol{u}$ or $\rho F_x u$. For the transformation of ρ we have

$$\rho = \frac{e}{V} = \gamma_u^{-1}\frac{e}{V_0} = \gamma_u^{-1}\rho_0 \tag{10.6}$$

Consequently

$$\frac{\phi}{V} = \gamma_u^{-1} u \left(\frac{f}{V}\right)_{0x} \tag{10.7}$$

$$g_x = \left(\frac{f}{V}\right)_x = \gamma_u^{-1}\left(\frac{f}{V}\right)_{0x} = g_{0x} \tag{10.8}$$

$$g_y = \left(\frac{f}{V}\right)_y = \left(\frac{f}{V}\right)_{0y} = g_{0y} \tag{10.9}$$

$$g_z = \left(\frac{f}{V}\right)_z = \left(\frac{f}{V}\right)_{0z} = g_{0z} \tag{10.10}$$

We derive immediately

$$(c^{-1}\phi V^{-1})^2 - g_x^2 - g_y^2 - g_z^2 = g_{0x}^2 = \text{const.} \tag{10.11}$$

thus confirming that

$$c^{-1}\phi V^{-1}, \quad g_x, \quad g_y, \quad g_z \tag{10.12}$$

are the components of an M-vector and that

$$c^{-1}\phi V^{-1}, \quad -ig_x, \quad -ig_y, \quad -ig_z \tag{10.13}$$

are the components of a cartesian 4-vector.

We shall now outline a more sophisticated approach through the Lagrangian and Hamiltonian. For the Lagrangian $\mathscr{L}$ of a charged particle in an electromagnetic field we replace (4.14) by

$$\mathscr{L} = -\gamma_u mc^2 - e\psi + eu_x A_x + eu_y A_y + eu_z A_z \tag{10.14}$$

where as usual

$$\gamma_u = (1 - u^2/c^2)^{\frac{1}{2}} \tag{10.15}$$

From (14) we derive

$$p_x = \frac{\partial \mathscr{L}}{\partial u_x} = \gamma_u^{-1} mu_x + eA_x, \quad p_y = \ldots, \quad p_z = \ldots \tag{10.16}$$

and consequently

$$mu_x = \gamma_u(p_x - eA_x), \quad mu_y = \ldots, \quad mu_z = \ldots \tag{10.17}$$

The total energy E is given by

$$\begin{aligned} E &= u_x p_x + u_y p_y + u_z p_z - \mathscr{L} \\ &= \gamma_u^{-1} mu^2 + \gamma_u mc^2 + e\psi \end{aligned} \tag{10.18}$$

or

$$E - e\psi = \gamma_u^{-1} mc^2 = \gamma_u^{-1} E_0 \tag{10.19}$$

Substituting (17) into (18) we obtain

$$E - e\psi = \gamma_u \{mc^2 + m^{-1}(p_x - eA_x)^2 + m^{-1}(p_y - eA_y)^2 + m^{-1}(p_z - eA_z)^2\} \tag{10.20}$$

If we multiply (19) by (20) the γ_u's disappear and we have

$$\left(\frac{E - e\psi}{c}\right)^2 = \left(\frac{E_0}{c}\right)^2 + (p_x - eA_x)^2 + (p_y - eA_y)^2 + (p_z - eA_z)^2 \tag{10.21}$$

Since the Hamiltonian $\mathscr{H}$ is the total energy E expressed as a function of x, y, z, p_x, p_y, p_z, we may rewrite (21) as

$$\left(\frac{\mathscr{H} - e\psi}{c}\right)^2 - (p_x - eA_x)^2 - (p_y - eA_y)^2 - (p_z - eA_z)^2 = \left(\frac{E_0}{c}\right)^2 \tag{10.22}$$

We recognize the components

$$(\mathscr{H} - e\psi)/c, \quad p_x - eA_x, \quad p_y - eA_y, \quad p_z - eA_z \tag{10.23}$$

of an M-vector of length E_0/c.

Differentiating (22) with respect to p_x and using (19) we obtain

$$\gamma_u^{-1} m \frac{\partial \mathscr{H}}{\partial p_x} = p_x - eA_x \tag{10.24}$$

so that

$$u_x = \frac{\partial \mathscr{H}}{\partial p_x} = \gamma_u m^{-1}(p_x - eA_x) \tag{10.25}$$

in agreement with (17). There are similar formulae for u_y and u_z.

Differentiating (22) with respect to x and using (19) we obtain

$$\gamma_u^{-1} m \frac{\partial \mathcal{H}}{\partial x} = \gamma_u^{-1} m e \frac{\partial \psi}{\partial x} - (p_x - eA_x) e \frac{\partial A_x}{\partial x}$$

$$- (p_y - eA_y) e \frac{\partial A_y}{\partial x} - (p_z - eA_z) e \frac{\partial A_z}{\partial x} \qquad (10.26)$$

so that by use of (17)

$$\frac{\partial \mathcal{H}}{\partial x} = e \frac{\partial \psi}{\partial t} - eu_x \frac{\partial A_x}{\partial x} - eu_y \frac{\partial A_y}{\partial x} - eu_z \frac{\partial A_z}{\partial x} \qquad (10.27)$$

The electrodynamic force acting on the charged body has an x-component

$$f_x = \frac{\mathrm{d}(\gamma_u^{-1} m u_x)}{\mathrm{d}t} = \frac{\mathrm{d}(p_x - eA_x)}{\mathrm{d}t}$$

$$= - \frac{\partial \mathcal{H}}{\partial x} - eu_x \frac{\partial A_x}{\partial x} - eu_y \frac{\partial A_x}{\partial y} - eu_z \frac{\partial A_x}{\partial z} \qquad (10.28)$$

Substituting (27) into (28) we obtain

$$f_x = - e \frac{\partial \psi}{\partial t} + eu_y \left(\frac{\partial A_y}{\partial x} - \frac{\partial A_x}{\partial y} \right) - eu_z \left(\frac{\partial A_x}{\partial z} - \frac{\partial A_z}{\partial x} \right)$$

$$= eE_x + eu_y B_z - eu_z B_y$$

$$= eE_x + e(\boldsymbol{u} \times \boldsymbol{B})_x \qquad (10.29)$$

There are similar formulae for f_y and f_z. Consequently

$$\boldsymbol{f} = e\boldsymbol{E} + e(\boldsymbol{u} \times \boldsymbol{B}) \qquad (10.30)$$

$$\boldsymbol{F} = \boldsymbol{E} + (\boldsymbol{u} \times \boldsymbol{B}) \qquad (10.31)$$

confirming the correctness of formula (2).

We can rewrite (22) in the form

$$\mathscr{H} - e\psi = mc^2 \left\{ 1 + \frac{(p_x - eA_x)^2 + (p_y - eA_y)^2 + (p_z - eA_z)^2}{m^2c^2} \right\}^{\frac{1}{2}} \tag{10.32}$$

to which the pre-relativistic approximation is

$$\mathscr{H} - e\psi = mc^2 + \frac{(p_x - eA_x)^2 + (p_y - eA_y)^2 + (p_z - eA_z)^2}{2m} \tag{10.33}$$

or on omitting the physically irrelevant constant term mc^2

$$\mathscr{H} = e\psi + \frac{(p_x - eA_x)^2 + (p_y - eA_y)^2 + (p_z - eA_z)^2}{2m} \tag{10.34}$$

The essential point confirmed by this lengthy algebra is just this. If we accept that the force on a stationary charge e is $e\boldsymbol{E}$, then the principle of relativity requires that the force on a moving charge must be $e\boldsymbol{F} = e(\boldsymbol{E} + \boldsymbol{u} \times \boldsymbol{B})$.

CHAPTER 11

Statistical Mechanics

THE following brief discussion of statistical mechanics will for the sake of simplicity be confined to a gaseous phase composed of one or several perfect monatomic gases. The model used of the molecules is structureless particles which can collide elastically but, apart from collisions, have no mutual interactions. Except at extremely low temperatures, with which we are not concerned, the three translational degrees of freedom of each molecule may be treated as classical (non-quantal).

According to pre-relativistic theory the fraction of molecules of mass m with coordinates in the ranges x to $x + \mathrm{d}x$, y to $y + \mathrm{d}y$, and z to $z + \mathrm{d}z$ and with momenta in the ranges P_x to $P_x + \mathrm{d}P_x$, P_y to $P_y + \mathrm{d}P_y$, and P_z to $P_z + \mathrm{d}P_z$ is at equilibrium equal to

$$\frac{\exp(-E/\Theta)\,\mathrm{d}P_x\,\mathrm{d}P_y\,\mathrm{d}P_z\,\mathrm{d}x\,\mathrm{d}y\,\mathrm{d}z}{\int\!\!\int\!\!\int\!\!\int\!\!\int\!\!\int \exp(-E/\Theta)\,\mathrm{d}P_x\,\mathrm{d}P_y\,\mathrm{d}P_z\,\mathrm{d}x\,\mathrm{d}y\,\mathrm{d}z} \tag{11.1}$$

where Θ has the same value for different kinds of molecules so that Θ defines a temperature scale.

The derivation of (1) requires only three assumptions. The first assumption is that the gas as a whole is at rest in the chosen frame of reference. The second assumption is the conservation of energy. The third assumption is the validity of Liouville's theorem. Since all three assumptions are as valid in special relativity as in pre-relativistic theory it follows that (1) is applicable according to special relativity. The only significant change introduced by special relativity is the form of the energy E. According to formula (3.20) we have

$$E^2 = E_0{}^2 + c^2P^2 = m^2c^4 + c^2P^2 \tag{11.2}$$

so that

$$\begin{aligned} E &= mc^2(1 + P^2/m^2c^2)^{\frac{1}{2}} \\ &= mc^2\{1 + (P_x^2 + P_y^2 + P_z^2)/m^2c^2\}^{\frac{1}{2}} \end{aligned} \tag{11.3}$$

as compared with the pre-relativistic approximation

$$E = mc^2 + P^2/2m = mc^2 + (P_x^2 + P_y^2 + P_z^2)/2m \tag{11.4}$$

Since Θ is independent of P we can determine Θ by consideration of small values of P when the numerator of (1) becomes

$$\exp(-E/kT)\,\mathrm{d}P_x\,\mathrm{d}P_y\,\mathrm{d}P_z\,\mathrm{d}x\,\mathrm{d}y\,\mathrm{d}z \tag{11.5}$$

where T is the usual thermodynamic temperature and k is Boltzmann's constant. The expression (1) then becomes

$$\frac{\exp(-E/kT)\,\mathrm{d}P_x\,\mathrm{d}P_y\,\mathrm{d}P_z\,\mathrm{d}x\,\mathrm{d}y\,\mathrm{d}z}{\iiiiiint \exp(-E/kT)\,\mathrm{d}P_x\,\mathrm{d}P_y\,\mathrm{d}P_z\,\mathrm{d}x\,\mathrm{d}y\,\mathrm{d}z} \tag{11.6}$$

In relativity theory there is no equipartition of energy, but there is equipartion of quantities like u_xP_x. For the mean value $\langle u_xP_x\rangle$ of u_xP_x is given by

$$\begin{aligned} \langle u_xP_x\rangle &= \frac{\iiint u_xP_x \exp(-E/kT)\,\mathrm{d}P_x\,\mathrm{d}P_y\,\mathrm{d}P_z}{\iiint \exp(-E/kT)\,\mathrm{d}P_x\,\mathrm{d}P_y\,\mathrm{d}P_z} \\ &= \frac{\iiint (\partial E/\partial P_x)\,P_x \exp(-E/kT)\,\mathrm{d}P_x\mathrm{d}P_y\mathrm{d}P_z}{\iiint \exp(-E/kT)\,\mathrm{d}P_x\,\mathrm{d}P_y\mathrm{d}P_z} \end{aligned} \tag{11.7}$$

On integration by parts of the numerator (7) becomes

$$\langle u_x P_x \rangle = \frac{\iiint kT \exp(-E/kT)\,\mathrm{d}P_x\,\mathrm{d}P_y\,\mathrm{d}P_z}{\int\!\!\int\!\!\int \exp(-E/kT)\,\mathrm{d}P_x\,\mathrm{d}P_y\,\mathrm{d}P_z} = kT \qquad (11.8)$$

Consequently for each monatomic gas in a mixture at equilibrium

$$\langle u_x P_x \rangle = \langle u_y P_y \rangle = \langle u_z P_z \rangle = kT \qquad (11.9)$$

In the pre-relativistic approximation $u_x P_x = mu_x^2$ which is twice the kinetic energy of motion in the x-direction. In special relativity $u_x P_x$ has no such simple significance. It is noteworthy that in the derivation of (9) we made no use of formula (3).

CHAPTER 12

Summary of Assumptions

We recall the basic assumption of Einstein's principle of special relativity as reformulated by Minkowski. The principle of special relativity requires that all the physical laws covered by this principle can be expressed in terms of M-vectors with real components or in terms of cartesian 4-vectors with some imaginary components.

This basic assumption has to be supplemented by further assumptions that several quantities are scalars in the four-dimensional world, that is to say their values are the same in all Galilean frames of reference. These quantities are the following:

1. The magnitude c of the speed of light: explicitly stated by Einstein.
2. The magnitude e of electric charge: explicitly stated by Sommerfeld.
3. The magnitude S of entropy: explicitly stated by Planck.
4. The permeability μ_0 of a vacuum: tacitly assumed but apparently not stated in the literature.
5. The permittivity ε_0 of a vacuum: this follows from (1) and (4) owing to the identity $\varepsilon_0\mu_0c^2 = 1$.

CHAPTER 13

Historical Synopsis

1904 Lorentz[1] in a paper entitled "Electromagnetic phenomena in a system moving with any velocity less than that of light" discovered the formulae, named by Poincaré the "Lorentz transformation group", which later became absorbed into special relativity theory.

1905 Poincaré[2] in a paper entitled "Sur la dynamique de l'électron" wrote "Il semble que cette impossibilité de démontrer le mouvement absolu soit une loi générale de la nature".

Poincaré[3] in a paper entitled "Sur la dynamique de l'électron" wrote "Il semble que cette impossibilité de mettre en évidence expérimentalement le mouvement de la Terre soit une loi générale de la Nature; nous sommes naturellement portés à admettre cette loi, que nous appellerons le *Postulat de Relativité* et à l'admettre sans restriction".

1906 Einstein[4] in a paper entitled "Zur Elektrodynamik bewegter Körper", independently of Lorentz and of Poincaré, wrote* "The same laws of electrodynamics and optics will be valid for all frames of reference for which the equations of mechanics hold good. We will raise this conjecture (the purport of which will hereafter be called the 'Principle of Relativity') to the status of a

*All English translations from the original German are by W. Perrett and G. B. Jeffery.

postulate, and also introduce another postulate, which is only apparently irreconcilable with the former, namely that light is always propagated in empty space with a definite velocity *c* which is independent of the state of motion of the emitting body." In the words of Pauli: "It was Einstein, finally, who in a way completed the basic formulation of this new discipline. His paper of 1905 was submitted at almost the same time as Poincaré's article and had been written without previous knowledge of Lorentz's paper of 1904. It includes not only all the essential results contained in the other two papers, but shows an entirely novel, and much more profound understanding of the whole problem." Actually Einstein's paper seems to contain a third postulate, namely that the magnitude of an electric charge is not affected by movement.

1907 Minkowski in a paper(5) entitled "Das Relativitätsprincip" and another paper(6) entitled "Raum und Zeit" reformulated the whole theory in terms of four-dimensional world vectors. This formulation is not only more elegant but also more powerful than the earlier formulation.

1908 Planck(7) in a paper entitled "Dynamik bewegter Systeme" revised and clarified the formulae for relativistic mechanics. By assuming that the entropy of a system is unchanged by motion, he also derived the formulae for relativistic thermodynamics.

1909 Lewis and Tolman(8) rederived the formulae of relativistic mechanics without making use of electrodynamics.

REFERENCES

1. Lorentz. *Proc. Acad. Sci. Amsterdam* **6**, 809 (1904).
2. Poincaré. *C. R. Acad. Sci., Paris* **140**, 1504 (1905).
3. Poincaré. *R. C. Circ. mat. Palermo* **21**, 129 (1906).
4. Einstein. *Ann. Phys. Lpz.* **17**, 891 (1905).
5. Minkowski. *Math. Ges. Göttingen*, 5 Nov. 1907; *Ann. Phys. Lpz.* **47**, 927 (1915).
6. Minkowski. *Cong. Sci. Köln*, 21 Sept. 1908; *Phys. Zeit.* **10**, 104 (1909).
7. Planck. *Ann. Phys. Lpz.* **26**, 1 (1908).
8. Lewis and Tolman. *Phil. Mag.* **18**, 510 (1909).

REFERENCES

[illegible]